I HAVE KNOWN
THE EYES ALREADY

A Research Memoir

Morgan Worthy

I Have Known the Eyes Already
A Research Memoir

ISBN-13: 978-1479232550
ISBN-10: 1479232556

INTRODUCTION

I am a retired psychologist. More than 40 years ago I started studying the relationship between eye color and behavior in humans and animals. In 1973, *Newsweek* had an article, *The Eyes Have It,* in their "Science" section, which was devoted to my research. That led to a lot of similar articles in other publications and interviews on radio and television. I was a guest on the show, *To Tell the Truth*. "Will the real Morgan Worthy please stand up?"

This memoir is an attempt to let the real Morgan Worthy stand up. The beginning theme will deal with my role as a researcher trying to study a topic, eye color and behavior, outside the mainstream of current psychological research. Prior to 1974, I studied humans and animals; after 1974, all my research was of animals. Here, I want to give a final overview of those research findings and tell what happens to a fellow who wanders off the beaten path of current research.

Then I want to tell, for the first time, about a personal research project that I undertook to understand why I kept injuring my right thumb. It took me 15 years to sort it out, but since 1991, I have not injured that thumb again. It all had to do with the curious fact that I am one of the few people who has been involved in three gun accidents. The first, and worst, happened more than 70 years ago; in fact, it happened Tuesday noon, the week before the Sunday morning attack on Pearl Harbor. Many lives were changed that week, and my life was no exception. I was five years old at the time.

Another theme I will try to touch on is the creative process. I will try to show how my particular quirks and knacks influenced my professional and personal research. In this area, I am especially interested in *aha moments*. Relevant to this theme is one of my earlier books, *Aha: A Puzzle Approach to Creative Thinking*. Will Shortz, now crossword puzzle editor at the New York Times, and earlier, editor at Games magazine, became interested in one type puzzle that I introduced in *Aha*. With his influence, that type puzzle became a fad. In Britain, the *Daily Express* called it a *Ditloid*. If interested, you can read in *Wikipedia* how they came up with that name.

My wife amuses herself by looking for faces in clouds. Recently, I told her, as a joke, that, as a psychologist, I look for clouds in faces. Only later did it occur to me that there is some truth in that and it goes back a long way in my personal history. My father, loving when sober, was dangerous when drinking. He also had very light blue eyes. I must have learned to study those eyes at an early age. A residual of that is that I seem to notice people with light eyes, even in black and white photographs.

One day, probably in early 1971, I was looking through a magazine dealing with (American) professional football. I noticed, once again, that there were many African-American players who had made it to this advanced level of skill and that they were not evenly distributed across all positions. As I neared the end of the magazine, I had the strange, vague, feeling of being reminded of some remote association. I had lingered on this page looking at a photograph of a white player with very light eyes. Then I had my *aha moment*: earlier in looking at the magazine I had stopped to look at another photograph of a white player with very light eyes, and in both cases the player was a quarterback. Now I recognized, consciously, what had unconsciously caused the vague feeling of remote association. Of course, it might have been a coincidence not worth remembering at all, but then again, I had learned, in military intelligence, to pay attention to even minimal bits of matching information.

Almost at once, I began to wonder if white players at different positions had different levels of average eye darkness and, if so, whether this rank order of positions was positively correlated to the rank order of positions based on percentage of African-Americans playing the position. When I later tested my speculations, the answer was "yes", on both counts. The two rank orders were positively and significantly correlated and both had quarterbacks at one extreme, with defensive backs at the other.

Defensive backs (especially those playing man-to-man) are much more dependent on immediate, quick reactions than are quarterbacks, who depend more on delayed, sudden reactions. Having already been thinking about the role of quick reactions in sports for several years, I jumped to the potential conclusion (i.e. hypothesis) that dark eyes are associated with the ability to make quick reactions. That started me thinking some more.

It occurred to me that eye darkness (not race or skin color) was the key dimension that could incorporate all the data. I thought in terms of *eye darkness* rather than *eye color* because, fortunately, I had been looking at black and white photographs in the magazine.

Also, it occurred to me that eye darkness, as a variable to study scientifically, had the advantage, unlike race, of retaining similar meaning across species. The more I thought about it, the more I thought of eye color, or eye darkness, as potentially important in scientific research.

Those insights led me to do archival studies of humans and animals related to eye darkness, and that led, in late 1973, to Eleanor Clift, a reporter at *Newsweek*, hearing about a book I was to publish in early 1974 and interviewing me for an article. Soon after the book was published, the publisher went out of business and the book was out of print and unavailable until I managed to have it reissued in 1999. Meanwhile, I was fortunate that some researchers, who are well qualified to study reaction times, decided that my idea was worth testing in a well-controlled laboratory setting.

A series of studies were done at Penn State University by Daniel Landers and his colleagues to test what has been called the "Worthy reactivity hypothesis." This is my idea that dark eyes are associated with quick reactions. (The hypothesis is not suggesting anything about you or anyone else as an individual.) After finding the hypothesis confirmed in seven straight studies using laboratory equipment designed to detect small differences in reaction time, they calculated that the chance that dark eyes are not associated with quick reactions is *less than one in ten million.* I can live with those odds of being wrong.

They demonstrated that the results were not related to differences in skin color. It is an eye-darkness phenomenon. Most of their studies involved comparing brown-eyed Caucasians with blue-eyed Caucasians. [These researchers were careful, as was I, not to confound race and eye darkness. However, at some point, we need to accept the fact that eye darkness is the critical dimension, and include in our human research a full-range measure of eye darkness that includes all people, regardless of race. There is already some evidence to suggest that eye darkness may be a better predictor, than is race, of risk for certain forms of hypertension.]

Partly because the differences between humans were small in absolute terms, I started in the 1970s to collect, mostly from field guides, published information on eye color for different species of land vertebrates. By the time this database, in its final form, was published in 2000, my wife and I had found published information on eye colors for 5,620 species of land vertebrates. Thousands were species of birds, hundreds were species of amphibians, reptiles or mammals. I need to make clear that my reactivity hypothesis is intended, now, to apply only to adult land vertebrates--not children, fish or invertebrates.

After comparing eye color information to behavioral information, it seems to me that the pattern holds across all classes of land vertebrates. One can see this by looking, first, at birds and bats. It is only the darkest-eyed families (mostly comprised of species with black or dark brown iris colors) that specialize in feeding on the wing in an open environment. That behavior is very dependent on speed and quick reactions. At the human level, that is analogous to outfielders in baseball; they, too, must have the speed, quick reactions, and developed skills to *catch flies* in an open environment.

At the other extreme, lightest-eyed, one finds herons. Their eye colors are mostly not dark at all, but yellowish, as are the eyes of families of frogs, cats, geckos and vipers. (These are the lightest-eyed large families in our database and come from all four classes of land vertebrates.) These animals are all hunters that lie-in-wait or slowly stalk prey before a sudden strike or pounce. All have some form of spring-loaded anatomy, such as folded neck, coiled tongue, or coiled body, that aids in making a sudden strike. At the human level, this is somewhat analogous to a slow-running quarterback in American football who, nevertheless, manages to be successful because of his ability and developed skill to just wait, with cocked arm, in a "pocket" of blockers, until the right moment to make a sudden strike downfield to an open receiver. Waiting, good timing and sudden release are all critical elements in the sequence.

It is easy enough to see in nature that yellow-eyed predators and black-eyed predators differ. Yellow-eyed predators use a tactic of *WAIT WITHOUT MOVING*. Black-eyed predators, such as those that feed on the wing, rely on a tactic of *MOVE WITHOUT WAITING*. Animals with eye darkness in the midrange between yellowish colors and dark brown or black (blue, green, gray, orange, red, hazel, light brown, brown) tend not to be skilled hunters, but, rather, rely more on finding immobile food (e.g. fruit, carrion, grubs, grass, eggs, ants, spiders). I have characterized this behavior as *self-paced*, or *CAN WAIT*. At least on the timing dimension, this is analogous in human sports to activities that are self-paced, such as pitching in baseball, shooting free throws in basketball, and the sports of golf and bowling.

[Land vertebrates that can hunt in total darkness tend to be dark-eyed and rely heavily on *KEEN senses other than vision*-such as *hearing* (e.g. Barn owls), *touch* (e.g. Boat-billed heron) or *smell* (e.g. pittas).]

To make sure that I was not "cherry-picking" my observations, I had twenty-one ornithologists make blind ratings of quick-versus-deliberate behavior for large families of birds. *Those ratings confirmed that, in birds, controlling for differences in size, light eyes were associated with deliberate behavior and dark eyes were associated with quick behavior.* Herons were rated as most deliberate and swifts received the highest ratings for quickness.

In addition to being very cat-like in hunting behavior, some herons rely on passive, MUST WAIT, defense. They use their streaked coloration to hide in reed beds with their bills pointed straight up. In contrast, swifts are very aerial and sometimes spend years aloft without landing. They use flight speed of almost 200 miles an hour and active, MUST NOT WAIT, reactions as their primary method of defense.

One implication of the above information is that a skilled predator will very often have an eye color that is yellowish or dark brown/black. However, the fact that a species has yellow or black eyes does not indicate that it is a skilled predator because many herbivores or omnivores have these eye colors also. The important thing to remember is that if the animal is a predator, knowing its eye color will allow you to predict its hunting strategy.

The pattern of eye color and behavior in land vertebrates is given in more detail in an earlier book, *Animal Eye Colors: Yellow-eyed Stalkers, Red-eyed Skulkers & Black-eyed Speedsters*. Descriptive statistics and brief analyses of differences between taxonomic families (within each order) are given for all orders of land vertebrates for which we have sufficient information on species eye colors. Included are analyses for two orders of amphibians, three orders of reptiles, five orders of mammals, and 15 orders of birds. The consistency across all four classes of land vertebrates demonstrates the general nature of the phenomenon.

To the degree that this memoir includes new information, or new terminology, it completes a trilogy that began many years ago with my first book, *Eye Color: A Key to Human and Animal Behavior*. As I learned with this attempt at a memoir, there are other, more personal, themes now ready to push aside my obsession with eye color. There are personal stories to tell and who knows how much time there is left to tell them.

In order to make the writing task easier, I used the device of edited transcripts of *conversations with my friend, John*. What I had not counted on was that the technique would have a force and direction of its own. It seemed almost automatically to tap into material that was personal or unconscious. The emotions aroused were real. The traumatic memories tapped into were also real and they seemed to force their way into the *conversations* such that it would not be easy to delete them without starting over. I had intended to write a research memoir about eye color and behavior, but, along the way, it evolved into a more personal memoir. However, that too, in its own way, is a research memoir.

A bibliography is provided for readers who have an interest in eye color research or want to read more about other topics that made their way into the *conversations.*

❖ **1**

MY FRIEND JOHN: All right, the recorder is on. This is your project. Get us started.

MORGAN WORTHY: The first thing that occurs to me is that you are often there when I go out for coffee and we have long conversations. That reminds me of two people, no longer alive, that I used to drink coffee with and talk about research. The first was my major professor in graduate school, Jack Wright. I was his graduate assistant. My first day there, after we talked in his office for about 15 minutes, he said, "Let's go to coffee; but before we do I'll make a deal with you. I won't call you 'Sir' if you will stop calling me 'Sir'." He understood; he had served in the military, too. We really hit it off well. He was from Oklahoma; his father was a Cherokee and his mother was a Norwegian immigrant.

A few years ago, long after he died, I learned from a relative that my great, great, great grandfather was a Native-American man and probably Cherokee, based on where they lived. Maybe my bonding with Jack was tribal. At any rate, Jack and I spent many hours over the next few years drinking coffee and talking about ideas. Once, he cut a class he was supposed to teach in order to keep talking and, once we just talked all night.

He was very clever at designing laboratory experiments. I learned a lot about research and a lot about life from him. He would stir his coffee and say, "The evenings, mornings, afternoons, I have measured out my life with coffee spoons."

MFJ: T.S. Eliot. So, we have formed a coffee-drinking bond. Do you think I may be part Cherokee?

MW: Oh, maybe so, but neither of us is going to get any casino money. I may as well tell you about the second person, Jim Dabbs. He was another psychologist who liked to drink coffee and talk about research ideas. He was a colleague for 25 years. Since his main research interest was testosterone and mine was eye color, we had fun swapping stories about what we were finding. Neither of us pointed out that the other was too far out of the mainstream of the area in which we had gotten our primary training--social psychology. Jim co-wrote a popular book on testosterone and behavior, *Heroes, Rogues, and Lovers* (Dabbs and Dabbs 2000). You may want to read it for insight into why you crow so much after winning a tennis match.

He encouraged me and agreed that eye color is an interesting variable. One of his last publications was an article he co-authored with Jon Bassett (Bassett and Dabbs 2001) reporting a study that tested one of the hypotheses from my 1974 book on eye color (See Worthy 1999, p 142). Based on some scattered facts, I had speculated that light-eyed people feel the effects of alcohol less readily than do dark-eyed people and as a result tend to drink more and become physically addicted. For their study, Bassett and Dabbs obtained information from more than 10,000 Caucasian male prison inmates and more than 1,800 Caucasian women who had answered a national survey. For both groups, individuals with light eyes had consumed significantly more alcohol than had individuals with dark eyes.

MFJ: Before we go on, let me show you that I was paying attention even before we turned on the recorder. I will not be distracted. Are there any more associations behind your sudden idea that the solution to your problem of writer's block might lie in recording our conversations for possible publication?

MW: I am glad you are determined to avoid being distracted. That is a good sign. If you want to do research, you need to be capable of creative obsession. I just have one other free association and you may not like it. Because I think of you as "my friend John," that may have tapped into my memories of reading how Milton Erickson used that phrase. He was probably the most skilled clinical hypnotist of the 20th Century. When he sensed resistance from a client, one way he might deflect it was to tell a carefully chosen story about a friend that he called "my friend John" (Erickson 1964/1980).

MFJ: Give me an example so I will know what you are talking about.

MW: My favorite example, and the only one I can think of now, was not in a therapy session, but an experience Erickson had one time when he was giving a lecture. I may not remember it exactly, but I can give you the general idea. This particular group was not attentive and some people in the audience were whispering among themselves. They were showing resistance by ignoring his lecture and, in return, he felt so irritated with them that it was difficult to continue. He solved the problem completely in the following manner. As a way to illustrate some point he was making in the lecture, he told them a story about a person he identified as "my friend John." He said that John was a turkey farmer and loved his work, except for the fact that the turkeys were so noisy it got on his nerves. It was always, "gobble, gobble." One evening he was trying to rest and the noise was so irritating that he suddenly jumped up, opened the back door, and screamed at the top of his voice, "You are all just a bunch of turkeys." And, of course, Erickson leaned toward the audience and glared at them as he ended the story in full voice. That is the "my friend John" technique. John, I consider you every bit as real and wise as Erickson's mythical friend, John.

MFJ: Oh, thanks a lot. I may want to hear more about this guy Erickson, but let me see if I understand the main points. You want to use me, your friend John, to help you overcome some internal resistance. Your writer's block may be based on anticipating indifference, or worse, from the audience you wish to reach.

MW: Very good. That is it and goes deeper than I realized myself. I had thought about my resistance as a conflict between writing and other things I want to do, like reading and playing more tennis. Your analysis proves the value of my not trying to do this alone. Thanks. You have earned the free coffee already.

Now, that surprised me. Rather than go any deeper into the personal or emotional right now, I feel a need to talk some more about eye color and the reactivity hypothesis. Ask me some question on that topic.

MFJ: I can do that. I have more questions about some of the things we have talked about already. They can wait, except for one. You have talked about your difficulty in writing. Sum up for me very briefly, if you will, how you view your writing style.

MW: I can sum it up with several basic points that I think are relevant. First, most academic people or writers I know say they did better in school on essay tests than on multiple-choice tests. I did better on multiple-choice tests. My recognition memory is much better than my recall memory.

Second, I first learned to do research and write reports when I worked in military intelligence in the Air Force. I was a communications analyst and worked more or less alone. I had all day to look for possible patterns, networks, or anomalies in whatever intercepted radio traffic had come in during the past day. At the end of the day, every day, I wrote a top-secret report that needed to be as brief and to the point as possible.

When I went to graduate school, soon after leaving the military, I was ahead of the game on certain intuitive aspects of doing good research, but my writing skills or habits were a different matter. After I had been there for some months, I was at a meeting of faculty and graduate students. The discussion topic of the day was different writing styles. I kept my mouth shut. Toward the end of the meeting, one of my professors, Marvin Shaw, said, "Well, I can tell you about Morgan's writing style. If he wants to say, 'The man went to town,' he just writes, 'Town'."

That sums it up as well as I know how. My natural writing style is to be brief and to jump from point to point (free association) rather than produce a logical sequence with smooth transitions. I think of my jumps from one association to the next as almost like those bullets used in graphics for a presentation.

In informal conversation I get excited and talk too much. The idea is that these recorded sessions, when transcribed, may hit a good midpoint between my writing style and my talking style.

Now, please, *how about eye color?* Ask me a question.

MFJ: I have a good one. What is a simple, easily understood, example that you can give to illustrate the reactivity hypothesis with animals?

MW: Cats and dogs are a good place to start because everyone is familiar with the differences in their responses to the same situation. If you are out in the yard with your pet and it sees a squirrel nearby, what it does next will probably depend on whether your pet is a cat or a dog. The immediate response of a cat is to freeze, then crouch and start to stalk in preparation for an ambush. The immediate response of *most* dogs is to run, without delay, toward the squirrel and chase it. One is a stalker; the other is a chaser and uses immediate, direct pursuit. The first responses of cats and most dogs on sighting prey are very different from each other. Only after the prey has come close to the waiting cat or the cat has slowly worked its way close to the prey, does the cat suddenly pounce.

The typical dog makes quick moves; the cat makes sudden moves. Understanding the difference between those two words, *quick* and *sudden*, is necessary to understand everything else we will talk about. *Quick* implies an immediate reaction; *sudden* implies an abrupt move after some delay. The origins of the two words make this plain. "Quick" means "swift, lively." "Sudden" means literally "to approach secretly" and comes from two Latin words that mean "secretly" and "to go." One way to remember it is *immediate quick* and *delayed sudden*.

Another way to state this is that one is *quick* and the other is *deliberate*. If we can agree that most dogs tend to be quick and most cats tend to be deliberate, we can then move on to differences in eye darkness between the two. The reactivity hypothesis is that *dark eyes are associated with quick responses and light eyes are associated with deliberate responses.* Using our example, we can predict that dogs are darker-eyed than cats. A simple way to get a measure of eye darkness is to say that only brown eyes and black eyes are considered *dark* and all others are considered *light*.

MFJ: So, in effect, you can use a 2-point scale, *light* or *dark*, which is quantified as 0 or 1. And in this example you apply that scale to get a score for each breed of cat or dog.

MW: Yes. I mostly used a 5-point scale, but it is easier to test some things with a 2-point scale.

MFJ: I have another question about your scoring. What if a breed or species has both a light color and a dark color listed?

MW: Score the breed or species according to the lightest color listed. If it says "light-brown," or any color other than brown or black, score it as "light." In effect, anything lighter than "brown" is scored as **0** on the 2-point scale. Use only eye colors for adults if a distinction is listed.

Here is a table showing the number of light-eyed and dark-eyed breeds of cats and dogs based on eye colors given in the *Simon & Shuster's Guide to Cats* (Pugnetti 1983) and *Guide to Dogs* (Pugnetti 1980). Dogs tend to be significantly darker-eyed than cats. Of the 27 breeds of domestic cat, none are dark-eyed. They are all in the range of yellow-amber-orange-blue-green. None are at the other end of the scale—black, dark brown, brown—that we are treating as dark-eyed. The same is true for cats in the wild. Look with me here at the database (Worthy 2000, p44). In the wildlife literature we found eye colors for 15 species of cat. All had yellow or yellowish eyes except for one, the Ocelot, and its eyes are reddish brown. So, for 27 breeds of domestic cat and 15 species of cats in the wild, using our 2-point scale of eye darkness, every one of them gets a score of **0**.

MFJ: I think I see several things you are getting at here. Cats are so consistently light-eyed that you can use them as a prime example for *deliberate,* as opposed to *quick.* In terms of your reactivity model, cats are the poster-child, as it were, for "Not Reactive." They eventually make sudden responses, but they do not react immediately to a new stimulus. *Sudden, but not quick.*

MW: Good man. I am glad you wanted to do this with me. I need to take a break. And I need to mention that from now on I am going to indicate on the tape any breaks in our conversations. I won't bother to indicate whether it was a short pause to get some more coffee, take a walk, or a longer break of a day or more. If I do ever try to publish our conversations, I may need to divide the transcript into something like chapters and it may help to know when in fact there were breaks. The other thing I will need to do is add references next to comments about published studies.

To respond to your comment before we took a break—yes, among domestic pets, cats serve as the best example to illustrate deliberate hunting behavior. Among mammals in the wild, the cat family is one of the best examples. If you expand the focus to include all land vertebrates, other candidates for your "Deliberate" or "Not Reactive" poster would include such animals as herons, frogs, and vipers. Let's just stay with cats, though, as our best example and get to your next comment or you will be scolding me again for jumping around too much.

Your "sudden as a cat" is good, but "quick as a dog" does not work as well because there is too much variability in reactivity or quickness among different breeds of dog. All cats that I am aware of respond to prey in a deliberate manner. For dogs we have to say, "most domestic dogs" respond in a quick manner—some breeds do not. Some breeds of dog are very cat-like in terms of immediate response to prey. I know you have spent some time in Germany so maybe you could say, as they do, *quick as a weasel.*

Your last point was about the value of being able to use my scales at different levels of analysis—breeds and species. That same simple approach (**0** to **1**) is used to get a crude but valuable measure of eye darkness at the level of individuals, sub-species or species. Field guides do not provide information on eye color at taxonomic levels higher than species, so what I do is aggregate species scores to each higher level. For instance, with the 2-point scale you can easily get the percentage of dark-eyed species in each subfamily, family, suborder, order, or class of animals. With a 5-point scale you calculate the average eye-darkness.

Maybe, when we come back, we should talk some about behavioral differences between different breeds of dog. I need to show you some more comparisons that get us a step past cats vs. dogs.

MFJ: Before too long I want us to go back and pick up the thread of the creative process. I want to know more about you as a person or a researcher.

MW: You and that thread that you keep losing or I lose for you. My free associations to *thread*: My mother was a seamstress and as a toddler I played with spools of thread beside the sewing machine as she worked. In high school, I took a ninth-grade vocational course that included work in the spinning room of the on-campus cotton mill. During college, I worked my way partly by working as a sweeper in the weave room of a textile mill. One summer I worked the second shift and one semester I worked the third shift. The noise in the weave room was very loud and many of the folks who had worked there for years had suffered a lot of hearing loss. You don't know where I am going do you? Studies show that for people who have followed a career in loud-noise occupations, such as aviation or textiles, the resulting hearing loss is greater for light-eyed people than for dark-eyed people. (Thomas et al. 1981; Kleinstein et al. 1984). So how is that for talking about the personal and still getting it around to eye color?

MFJ: Show off. You were on a roll. Are you angry with me for wanting you to talk more about yourself?

MW: I don't think so. I was just having fun, but I agree my response did have an edge to it. If I had to guess, it would be that I want to talk about myself and about the process. However, if it turns out to be too much about me or I boast too much or whine too much, I want to be able to blame it on you. Take it as a compliment. If I had more self-awareness, I would not need you. I get the last word. Let's stop for the day.

❖2

MFJ: **I saw your light on and here I am. Consider us in session.**

MW: Good. I am glad you came on over. Anytime the light is on at this outside office, you will know that I couldn't sleep and came out here to work or read. Since you say you sleep as irregularly as I do, we may do some of these sessions in the wee hours of the morning.

MFJ: That is OK with me.

MW: Years ago, when I was first studying eye color, I would only sleep three or four hours a night so I could keep working on it. The problem is, since we started these sessions, dialogues continue in my head and I am getting very little sleep. That is another compliment. I am excited about what we are trying to do. However, so many ideas are flooding in and I am losing so much sleep, I have started to worry about it. Sometimes in the past I have dropped things because the excitement wrecked my sleep. I stopped playing poker for that reason. I do not intend to stop these sessions, but the very process brings up more personal or emotional issues than I expected, so I need to somehow defuse that.

MFJ: I think we need to put off talking about dogs a bit. How is that for being perceptive? Would I make a good psychologist?

MW: You are already on your way. Let us do this. I want just to mention some of the things that keep jumping into my mind and must be relevant, though I am only now starting to see any possible connection to eye color or to me as a researcher. What if I just mention a few of those now and we can come back to them later? Or am I pushing you out of your role of being the one who asks the questions?

MFJ: Not to worry. That sounds all right to me. I can defend my role. You won't bully me too much because you know I will more than pay you back on the tennis court.

MW: Oh, yes, tennis. That is one of the things I wanted to mention. I should have told you earlier that one reason I picked you for this is that every time we play as partners, I play my best tennis. When we are opponents, I play my worst tennis. That says something about the bond. During our last session, you may remember, I got agitated and ended up talking about noise and hearing loss.

MFJ: I remember.

MW: Well, only later did I realize where that came from and it was so obvious that it surprises me I could be so unaware. I have very bad allergies and for three days my ears have been plugged. I have not been able to hear as well as usual and noises have bothered me. Because this reaction is worse than usual, I have started to worry that I may follow my mother and her father, who both had serious hearing loss in later life. Now, more bullets of association:

- I have been remembering my father's eyes. I remember the dangerous, or piercing, look in his light-blue eyes when he was drinking and I have wondered if that caused me to notice eye color more than do most people.

- We spent time talking about sudden responses and how they are different from quick responses. One definition for "sudden" is *unforeseen*. It occurred to me that if I am going to tell you about myself, I need to tell you that I am the only person I know of who was involved in three gun accidents as a child. I was never the one holding the gun, but three guns went off unexpectedly in my presence. I think that changed my personality in some ways that I only realized later. And some of those ways influenced my strengths and weaknesses as a researcher.

- When I was a graduate student, one of my professors, Herb Kimmel, said one day in exasperation, "Morgan, you are always apologizing." I was hurt and did not know what he was talking about. Then I started listening to myself and realized that he was right. I had all these subtle ways of saying things that were defensive in nature or said in advance that I may be wrong. I tried to change that habit and made some progress, but it was subtle and engrained. I guess it was and is my way of trying to disarm people. In my unconscious mind, based on early experience, that metaphor "disarm" is on target. It does not take much to trigger my defensiveness. If all my use of metaphors seems like overkill or silly, just know that somewhere in here I have probably buried an apology or two.

Thanks for listening. I think that gets into the mix things I needed to say. That is a relief. We can come back to them later. Now we can move on to dogs if you are ready.

MFJ: I notice that you list these very emotional personal experiences and say let us move on, which does not allow or invite any comment from me. You would probably say that that is just another not too subtle way of defending yourself. My reading of it is that this process, to your surprise, has activated an earlier voice and you are wise to let it emerge, or not, slowly. What about eye color and dogs?

MW: "Earlier voice," you said. Now that is an interesting choice of words. I did, as a child, go through a period at ages five and six when I did more or less lose my voice. A failure to speak up for good or ill became an engrained habit. We can talk about that later. Let's go to the dogs. Is there anything you want to ask about eye color before we consider dogs?

MFJ: Yes. What causes eyes to appear brown or black?

MW: The amount of a black-brown pigment, eumelanin, in the iris is the main determinant of eye color. If there is a high enough concentration of eumelanin, the eye will appear brown. If the concentration is very high, the eye will appear black. If the particles of melanin are very small, a light-scattering effect will cause the iris to appear blue (for the same reason that the sky appears blue). Eye color is also determined by the amount of a yellow-red pigment, pheomelanin, in the iris. There are other factors involved, but that is the basic difference between dark eyes and light eyes. If you like, I can give you a recent article (Borteletti et al. 2003) that discusses various other factors that can influence iris color.

MFJ: What about the physiological links between eye color and behavior?

MW: I deal with the externals and not the internals. However, if you want to explore the internals I can tell you enough to get you started. Amount of melanin in the iris is correlated with amount of melanin in the inner ear (Bonnaccorsi 1965) and with amount or distribution of neuromelanin in the central nervous system (Happy and Collins 1972). In terms of the link to motor behavior, it is perhaps significant that neuromelanin can function as a semiconductor (McGinness et al. 1974). Eye color is polygenic and the specific genetic causes are still being sorted out (Zhu et al. 2004). I just use eye color or eye darkness as a marker variable that is external and easily observed. In fact, eye color was used as a marker variable in many of the early studies of genetics.

MFJ: Thanks. I am now ready to go to the dogs with you and it looks like my puns will have to be as corny as yours.

MW: Most dogs react to prey by immediately giving chase. One group of dogs, though, employ an initial response to prey that is very much like the initial response of cats. Pointers and setters, like cats, freeze when they first sense prey nearby. Pointers adopt a standing pose and setters crouch. In regards to this initial response, I think any fair observer would grant that pointers and setters are more deliberate or cat-like than are other dogs. If the reactivity hypothesis is correct, those breeds (all are often just referred to as *Pointers*) should be less likely than other breeds to have dark eyes. That is, indeed, the case. Whereas 70% of other breeds are dark-eyed, only 28% of the pointer or setter breeds are dark-eyed. A difference that large, given the sample sizes, could occur by chance less than one time in a thousand.

Consider the two findings. Cats are lighter-eyed than dogs, and pointers and setters are lighter-eyed than other dogs. In my view, a pattern starts to emerge. Because pointers are cat-like in both eye color and the critical behavior considered—first response to prey—we can suspect that cat-like hunting behavior and light eyes are linked.

Pointers are bred for "freezing" as first response to prey; hounds are bred to track and chase prey; terriers are bred for not only chasing the prey, but for following it into burrow or den— which requires a high level of persistence and courage. *Simon &*

Shuster's Guide to Dogs (Pugnetti 1980) uses a symbol to indicate adaptation for each of those three behaviors.

There is a progression. Fifty-five per cent of pointer breeds have yellowish eyes; for hounds, it is only 10%, and there is no breed of terrier that has yellowish eyes. Yellow eyes seem to be associated with hesitation or freezing behavior, which is good for animals that stalk. Hesitation would tend to be a liability for animals that hunt by means of direct pursuit. And that would be especially true for terriers, which are expected to pursue the prey into its den.

You know, I said earlier that we should follow the Germans and say "quick as a weasel." Weasels and terriers are very much alike in being quick, fearless in pursuing prey into its den, and being very dark-eyed. Dark-eyed animals show active courage; light-eyed animals that freeze when predators are near show passive courage.

One of the main things to remember, though, from our talking about cats and dogs, is that predators that depend a lot on freezing, ambush, lying-in-wait, stalking, or any other form of surprise to take prey will not only be light-eyed, but most likely will have yellowish eyes. I know we have only covered three examples so far—domestic cats, cats in the wild, and dogs that point or set— but the same pattern is seen with all classes of land vertebrates. Any type predator that uses surprise to ambush prey (in less than total darkness) tends to have yellowish eyes. That can be noted by anyone who cares to look within various orders or sub-orders of animals: frogs, snakes, lizards, crocodilians, carnivores, primates, raptors, owls, heron-like birds, and various other orders of birds.

No one can deny that statement, but they can ignore it. Given human history, people of good will are now reluctant to acknowledge any evidence that pigmentation can be related to behavior. We seem always to go from one extreme to the other.

Make a list sometime of all the families of predators (that hunt in less than total darkness) that rely heavily on some form of surprise to ambush prey, and then look at the database of eye colors. You will be surprised at how many of the species in those families have yellowish eyes. You cannot, though, just look for yellow-eyed species and expect them to be ambush predators. Many families of herbivorous animals have yellow-eyed species, black-eyed species, and species with eye colors in between. They tend to average out somewhere near the midrange of the eye darkness scale. Later on we can talk about animals that mostly eat vegetation, but by just concentrating for now on predators it is easier for me to introduce the reactivity hypothesis.

Let me mention, in passing, that predators that feed in total darkness tend to be dark-eyed; they use *KEEN* senses other than vision to locate prey. Good examples include the Barn Owl and the Boat-billed Heron. While I am talking about night creatures, perhaps I should go ahead and tell you that the Possum has no visible iris.

By the way, this may be a good time to mention that there are at least two good ways to test the model I am suggesting. One is to ask, "Are animals that are unrelated, but similar in the critical behavior, also similar in eye color?" A different approach is to ask, "Are animals that are closely related (such as different families in the same order), but different in behavior, also different in eye color?"

Of course, if one has measures of behavior, it is probably better to turn those statements around and think in terms of eye color as a predictor of the behavioral measure. The fact that cats, though not closely related to herons, are similar to them in behavior and eye color contributes to answering the first question. That the cat family, closely related to the weasel family (same order), differs from weasels in eye color and behavior in a meaningful way, contributes to answering the second question. Cats go quietly in order to surprise prey; weasels make a lot of noise in order to flush prey that they can chase, even into its den, if necessary.

We should stop now. I keep wanting to jump on ahead.

MFJ: Just so it is not on mine.

BREAK

MFJ: Let me see if I can remember the questions I had on my mind when we stopped last night. We were talking about dogs being more diverse than cats. Based on what you showed me, when talking about deliberate vs. quick in domestic animals, it might be better to say "cats vs. terriers" rather than "cats vs. dogs."

MW: Yes, to use your term, terriers are the poster-child for reactivity among domestic animals.

MFJ: Are there studies of terrier behavior in the literature?

MW: Yes, I know of some and there are probably others that I do not know about. Several compared terriers to beagles. Beagles are listed as having eye colors that are hazel or chestnut. Their eyes are darker than those of most pointers, but lighter than those of terriers. I can tell you what one person, James (1951), said after comparing the two breeds in controlled situations. The terriers were more "active" and came running forward to meet the observer. The beagles would hold back and if cornered would assume a passive defensive attitude. James characterized the terriers as definitely less inhibited than the beagles. Those differences are consistent with the hypothesis.

Helen Mahut (1958) did a study in Canada in which she compared ten breeds of dog on response to novel stimuli and categorized the behavior as "fearless" or "fearful," depending on how bold or inhibited the dogs were in their responses. I no longer remember the particular breeds, but when I checked the eye colors, the most fearless dogs were also the ones with the darkest eyes.

MFJ: If I remember correctly, dogs are related to or descended from wolves and pictures I have seen of wolves indicate that they tend to be light-eyed. Has anyone compared wolves and dogs on behavior?

MW: Well, like on many of these things, my references may be dated, but you can check the *Citation Index* to see if there is anything more recent. Asdell (1966) described wolves as being cautious, cowardly and fearful of novel stimuli. They pursue prey in a circling or zigzag manner in order to set up an ambush. That is not direct pursuit as is seen in terriers or weasels. Nor is it as non-reactive as the behavior of cats and pointers. Because wolves are lighter eyed than most dogs it is significant that Asdell also reported that wolf-dog hybrids exhibit "passive defense reactions" more than do most dogs. That fits with a topic I don't think we have talked about yet—differences between dark-eyed and light-eyed animals when they are considered as prey rather than predator. Light-eyed animals, domestic or wild, more often exhibit passive ways of escape than do dark-eyed animals.

I think that is all we need to talk about cats and dogs. I base that on the fact that I don't know anything else. We have spent a lot of time on the subject because you, like most people, are interested in pets and have observed them yourself. Another reason is that there is much information available on them already.

That is the archival approach: first see if data is already available in some archive that can be used to answer your research question. My guess is that people who do not like to even start with that approach, think of it as looking for your lost car keys under the lamp post because that is where the light is good. I understand; my training was in laboratory research. That approach, though, is expensive in time and money and forces you to focus too narrowly when you are trying to get a glimpse of the big picture.

Also, you have to ask, 'What is *my* particular knack?' I think mine is to recognize the vague outlines of a pattern contained in scattered data points. I say with pride that my work is a mile wide and an inch deep. Actually, the reactivity hypothesis has been explored in depth in a few areas such as simple reaction time and the bird and bat behavior of feeding on the wing. Even so, the strength of the hypothesis is that it applies so widely. Some fortunate observer had to be the first person to recognize this broad, intriguing, pattern in nature. I was lucky enough to be that person.

MFJ: It is more likely that the pattern has been noticed and commented on before without leading to an obsession. Darwin may have said something about it 150 years ago in a footnote.

MW: I deserved that. Now that I have been exposed as a legend in my own mind, we should stop for a while. John, you are such a good listener, I get carried away with the topic. As you can tell, I am *intrigued with the phenomenon*. We need to break; maybe next time we can start with that expression.

MFJ: When you get carried away with what you are talking about are the times I consider most worthwhile and enjoyable. That seems more like the real you I see on the tennis court or at the beach. And then you find some way to apologize. But, that is no problem. I am not bothered by displays of false modesty. I looked forward to this session today. It is interesting how

quickly a habit develops. Our not doing this yesterday made me feel like something was missing. So you have your hearing back and are ready to go? You were going to tell me more about being intrigued with the phenomenon.

MW: Well, there is a story behind the expression. Dan Levinson, a psychologist at Yale who is well known for his research into stages of adult development, gave a free lecture one evening at my university. The next day he offered a paid workshop. During the workshop, 30 or 40 minutes after it had started, a woman slipped into the back of the room. To her surprise, Dr. Levinson stopped what he was doing, got his list of the 40 or so people who had registered for the workshop, and asked her name. Trapped, she admitted that she had hoped to crash the workshop and get the benefit without paying. He told her she could stay but only if she agreed to come up front, be interviewed by him about her "crashing" experiences, and give honest answers to his questions. She agreed. That interview became the central focus of the day and we saw a master interviewer at work. From time to time, he would turn to us and apologize, "I know this is taking a lot of time, but I am just *intrigued with the phenomenon.*" I identify with that spirit and like the expression.

MFJ: I understand how fortunate you were to see him exercise his special skill exploring a topic that was fresh and exciting for him. From the way you described it, it sounded as though he was both excited and apologetic at the same time. No wonder

you identified with him. Which reminds me. Why do you say things like "in graduate school" or "my university" without naming them?

MW: I will answer that in just a minute, but it is very important to me to show you something first. It is a quotation from an article published in *Biological Psychology* (Hale et al.1980). I know you like me, but I do not know yet whether you think of my work as science or pseudo-science. Before I tell you anything more about myself, I want to tell you about this independent research done at Pennsylvania State University by people I have never met. They tested the reactivity hypothesis with human subjects by studying eye color and reaction time in a laboratory setting. They first found that simple reaction time is not related to skin color, but it is related to eye color. They found that dark-eyed blacks and dark-eyed whites have faster reaction times than do light-eyed whites. They then focused just on comparing dark-eyed Caucasians to light-eyed Caucasians on how quickly they could react to a visual or auditory stimulus. They did a number of well-controlled laboratory studies, and then did a meta-analysis of all those studies. Read this quotation which reports the results:

> Thus, the findings across studies have consistently shown that dark-eyed subjects have shorter pre-motor time and simple RT latencies than light-eyed subjects. Considering that Worthy's hypothesis has been experimentally tested seven times with seven different samples ... a combined

probability value would more accurately reflect the reliability of the eye color phenomenon. Using a z-transformation procedure ... a z value was obtained that could not occur by chance any more than one time in 10 million. Worthy's hypothesis, therefore, reliably predicts RT differences between eye color groups from one study to the next (Hale, et al. 1980, p. 61).

I can live with a probability of one in ten million that my hypothesis is wrong. I wanted you to read that in order to make it clear that the association between dark eyes and quick reactions is very well established in humans. Is it likely that dark eyes would be associated with quick reactions in one species, humans, and not in other creatures? If you think about it that way, it should not be a great surprise that the same relationship is evident when you look at animal behavior. It is only because the nature/nurture wars have left us thinking of human behavior as determined by culture and animal behavior as determined by genes that it seems like such a leap to go from human behavior to animal behavior.

MFJ: I want to read the whole article. It shows what you can do with laboratory research. The probability level would seem to remove any doubt that eye color is related to quickness. Were the differences large in absolute terms?

MW: No. They were not. Their studies were designed to answer primarily whether or not there was, indeed, a reliable phenomenon. Their studies answered that conclusively as far as I am concerned. I had reached the same conclusion by studying performance records of professional and college athletes. Even small differences in the general population can matter when looking at a heavily selected group like professional athletes.

My archival research of athletic performance led to their laboratory research. I gladly give them credit for providing the most irrefutable evidence for the reactivity hypothesis as it applies to humans. More than 30 years ago I stopped doing human research myself. I wanted to see the big picture, and any differences within one species, such as humans, are always going to be small in absolute terms.

For what I wanted to do, it made more sense to look at differences between species, between subfamilies, families, suborders, orders, and classes. I have mostly studied differences between families of animals (Worthy 1978; Worthy 2000). In the first of those publications, I used a 2-point scale to just nail down that there was a reliable phenomenon. After doing some content analyses, I then asked 100 ornithologists to make blind ratings of large families of birds on "quick-versus-deliberate" behavior related to flight, feeding, and escape. Twenty-one agreed to do so. Some left out those families with which they were not very familiar.

I included in the analysis all large families of birds for which at least 15 ornithologists had made ratings. When size was partialed out, the eye-darkness measure and the combined behavioral measures correlated .56 [d.f.=33, p < .001]. As you probably know, John, that means that differences in eye-darkness, even using a two-point scale, accounted for about 31% of the rated differences in quick-versus-deliberate behavior. That is not trivial. The family of birds that was rated most deliberate was herons; the family of birds that was rated quickest was swifts. Whereas the reaction time differences with humans were small in absolute terms, in this study of birds, the behavioral differences were large.

MFJ: You make good points and as usual you seem to do so in a defensive frame of mind. It is as though you expect to be attacked.

MW: I am vigilant. I blame that and my sometimes being socially clueless to the residuals of having had an alcoholic parent. That and guns that kept wanting to go off in my presence left me vigilant and jumpy. That has a down side, of course. It also has an up side. Being vigilant, I may sense a subtle cue and jump to a conclusion. Some times those conclusions prove to be correct and valuable insights. Some times not.

MFJ: I am intrigued with hearing you say more about what I've heard you describe as using your *knacks* and *quirks* to be more creative, but you were going to give me, and the mythical reader of this, some geographical context.

MW: "Mythical reader" – that is you, great grandchildren, and if I am lucky, one of you will be a comparative psychologist looking for a dissertation topic.

MFJ: Or maybe a history major who specializes in "pseudo-science of the 20ᵗʰ Century." I still have not heard why you do not identify your history more specifically as to location or affiliation.

MW: I grew up in South Carolina. The Worthy family came to Chester County in the 1700s. The Morgan family (my mother's family) came to Oconee County in the 1700s. I started school in Pickens County. I graduated from Parker High School in Greenville County. I attended North Greenville College, earned credits through the University of Maryland overseas program while I was in the Air Force and graduated with a B.A. from Furman University. I earned M.A. and Ph.D. degrees at the University of Florida in social psychology. I did a year's post-doctoral traineeship in clinical psychology at the Veterans Hospital in Hampton, Virginia. In 1966 I took a position at Georgia State University in Atlanta, where I worked first in the Psychology Department and then in the Counseling Center. My academic career, 27 years, was spent at Georgia State.

As a teenager, I worked three summers in Indiana. During my four years in the Air Force I was stationed first in Texas and California and then for the last two and a half years I was stationed in Germany.

To answer the "why" part of your question: I hope any potential reader or researcher will judge my ideas by looking closely at nature itself and judge the potential value of the reactivity hypothesis by that and not by who I am or where I live.

MFJ: There are so many ways to follow up on that and yet I do not see how any of them get us anywhere. People are free to use whatever clues they choose to get an idea of who we are and there is nothing we can do about it. We just make ourselves look bad when we fight it or try to prevent it.

MW: You are right on all counts. And thank you for being kind enough to express it as, "we just make ourselves look bad." "You have this defensive quirk" might have been more difficult for me to hear. I tried to forget something that one of my graduate professors told us. Because he had taught at different universities in different parts of the country, he thought his rejection rate from journals depended to some degree on geography. However, it is just too tempting to think, "That is my excuse and I am sticking to it."

Why don't you ask me a question on some other topic? *What about eye color?*

MFJ: "That is my story and I am sticking to it." You got that punch line from Alex Hawkins, the pro football player, who said his wife caught him sneaking into the house at dawn. He claimed he had spent the night in the tent in the back yard— not knowing she had taken in the tent the day before. I will have to stop soon. The next time we could talk about eye color, your other research, creativity, or you can free associate the way you do when we get into our semi-pseudo-quasi-psychotherapy routine. Before you have a chance to take that as a criticism, let me say, I enjoy the pseudo peer counseling we

do here and at the coffee shop. There is no way I will let you mention, on this tape, having been involved in three gun accidents as a child and not tell me about that before we finish. One day, when we are done with these sessions, we can start over with you as the counselor and look at my ideas and expose all my quirks and knacks.

MW: Thanks, John. This whole process started because I have been stuck for nine years unable to find a good way to take, in writing, my last bite of the eye color apple (as I notice that I am vaguely looking at the logo on my Imac.) Only this very day did it occur to me that the nine years of feeling blocked followed indirectly from something having to do with guns. I was very surprised and a little disappointed with myself that day I mentioned the gun accidents. I can see the relevance now. In my terms, my unconscious mind knew that information was relevant and somehow prompted me to drop it into our conversations.

Have a good weekend. I will see you at the tennis court.

❖ 3

MW: John, it feels strange to be back at this again after a day away from it. And meeting you in a different office also feels strange. By the time we finish this, we may have talked and recorded in a lot of different places—especially if we try to record any of our conversations while we are down at Cocoa Beach in May. On the other hand, we could really work at this and try to finish before we go to the beach.

MFJ: Things like that you will have to decide. Because our sessions are focused on you and your work, I will try to match your pace rather than set the pace myself. When we focus on me I will want to set the pace. I can see that this process taps into personal or emotional areas and I will want to feel free to slow it down or speed it up.

MW: Perceptive. I once heard that the true mark of humility is to see the needs of others as they see them.

MFJ: Speaking of which, thanks for letting me take home the series of charts and tables. I know it is not the way you had it

planned. **They have already made me think a lot and if it is alright with you, I want to keep them for some days before we talk about them again. My plan is to go back to my animal behavior books to see how what I read there relates to your theory. That way, we can compare notes. In the meantime, we can explore other things without feeling pressured to get back to eye color of animals. I will still be looking at animal eye color whenever I have time, but it will be without you telling me what to see or giving me examples; I will look for my own examples and we can compare notes. That is what will be the most fun for me, and before we started you said I should work with you to make it fun for both of us. What do you say?**

MW: You have a deal, even if you do use my own words to trap me. For a while, if we talk any at all about eye color, it will be about humans. Then, to use your expression, we will pick up the thread of animal behavior when we are both loaded for bear. The truth be told, these conversations have brought up so many memories, not all pleasant, that I welcome the chance to spill some of them out, bullet-style if need be, or more like free association that helps to empty some of the internal garbage.

MFJ: **I thought you guys called that *sharing*. By the time we finish with each other, we can say we built a bond by sharing garbage. Anything from the weekend you want to share before I dig deep into your early years?**

MW: Yes. I want to talk about a poem, *The Love Song of J. Alfred Prufrock* that T.S. Eliot wrote in 1917. That poem is in a book of poems (Eliot 1930) that I keep on my bedside table. Jack Wright used to quote lines from it. It has been my favorite poem for almost fifty years. If you had asked me a few days ago why I was so drawn to it, I would have said that I didn't know. I might have then speculated about how it reminded me of Jack or certain lines that I tend to remember. If you had then asked me whether it mentioned eyes at all, I think I would have answered honestly, "No, not that I recall." Yesterday, I picked up that book, opened it and saw lines that jumped out at me.

The paragraph begins:

And I have known the eyes already, known them all

He goes on to say that he is talking about eyes that leave you feeling like you are *sprawling on a pin* and *wriggling on a wall*.

From there, he ends the paragraph with these lines:

Then how should I begin
To spit out all the butt-ends of my days and ways?
And how should I presume?

I think those lines about eyes must have struck a nerve each time I read them. I recognized the feeling, felt connected to it and then, never recalled it afterward. That is perhaps like having a series of hypnotic experiences or dreams that one does not consciously recall. You may remember that I referred to my daddy's eyes as *dangerous* and *piercing*. One day I remember well, my dad was drinking, angry, and walked around all evening with a butcher knife in his hand. Without getting too Freudian about it, I would say that that evening gave me good training in being vigilant.

The last lines capture for me how difficult it is to know how to start telling or writing about the *butt-ends of my days and ways*, even if I think doing so would have value to me or someone else. The last line even captures my feelings of apology or, as you suggest, false modesty.

MFJ: Is there anything else about that poem you want to mention?

MW: There are some lines that I think may be meaningful to many of us whose early experiences caused us not to learn certain social norms, skills or use of words in exactly the same way as other people. Prufrock indicates the impact of being told:

That is not it at all,
That is not what I meant at all.

Some people just seem to know how to navigate socially at low cost. And then, some of us seem often to be surprised by a response we had not expected. That is, socially, we miss some cue, fail to anticipate some response or, in general, feel that we get surprised too often for that to be "normal." It is not serious enough to disable us socially and others may not see us that way at all. From what I read, many adults who spent time in a family with an alcoholic parent see that as a part of their personal make-up—a failure to feel at home with all the little unspoken social norms. I think of that as one of my quirks.

Even if I am right about the behavior, there are other possible causes of it. Traumas, like the gun accidents I have mentioned, may have left me a little out of alignment, like a car that has hit a curb. I would also say that there could be some genetic component involved. Research has indicated that a sub-set of blue-eyed children, especially little blue-eyed boys, are more likely to be inhibited or shy than other children (Rosenberg and Kagen 1987; Coplan et al. 1998). If I were shy from birth, that tendency would have influenced how I dealt with environmental events and their impact on me. None of us will ever be able to track down all the genetic and environmental influences that made us the unique individuals that we are. However, if we think about it, we do know where our quirks and knacks lie. Knacks that we have developed or strengthened and applied to our own creative endeavors may have followed from particular quirks.

MFJ: I think you are losing me. Define *quirks* and *knacks*. Or tell me where I can read about this idea.

MW: I am not aware of any place you can read about it. It is just a part of my attempt to think *psycho-logically*. Some of it I thought of as I was walking over here. I tend to try new terminology very readily. That is one of my quirks that is a weakness in some research situations and a strength in others. That tendency has been an asset in trying to go "a mile wide" and look at the animal kingdom as a whole. It would be a weakness if instead I wanted to go a mile deep. In that case, I would need to rely more on tight, operational definitions.

MFJ: So your logic may be psycho, but you avoid getting stuck or pinned down by always changing your terminology. If you can not disarm your critics, you can at least make yourself a moving target. How is my analysis?

MW: Too damn good for my comfort, but it misses the main point. Or at least the point I was trying to make, which is that knowing your own quirks and knacks helps you to decide where and how you can make a creative contribution.

My father is a good example. He started many small businesses having to do with the repair of automobiles. He had a paint shop, a body and fender shop, a radiator shop and others along the way. He had a knack for getting businesses off the ground, but he was like an airplane pilot whose skill and training is all in takeoffs. Crashes tend to occur to entrepreneurs who have the quirk of engaging in month-long alcoholic binges.

On the other hand, my father had a knack for invention. The only idea he patented was a filter to be attached to a paint-gun. That eliminated one messy step in the old procedure. He had first gotten the idea when he had the job of painting some jeeps while he was in the army. After the war, he had a shop for painting cars, which he ran out of our detached garage. That reminded him of his idea and he carried through to have it patented. He sold his rights and took off on another extended holiday. That was more than 50 years ago and if you need a really good filter for your paint gun, you can still buy a "Worthy Filter." [Actually, it is a "Worthy Strainer."]

MFJ: I will have to check that out on the web. If I understand the import of that story, your father's failed businesses followed from his not seeing or accepting that his life was marked by discontinuity. If he had accepted that, he could have made his efforts in areas, like the invention, that did not require the same continuity as does running a business.

MW: Exactly. And all of us have to learn such things about ourselves and use that knowledge or we pay a price. I equate it to a key rule I learned in whitewater canoeing and kayaking: *lean downstream.* And now, I need to honor your request to define *quirks* and *knacks.* Let us just see how those terms are defined in my American Heritage dictionary.

> *QUIRK: A peculiarity of behavior that eludes prediction or suppression.*
>
> *KNACK: A specific talent for something, especially one that is difficult to explain or teach.*

MFJ: Looking at those two definitions, I can see how they go together and why you used some of the illustrations you did. One knack or talent you want me to develop is to recognize my quirks and be creative in making decisions based on that information.

MW: Yes. It is a variation on the cliché that is used in vocational counseling: *First, decide what it is that you like to do well and then, find someone to pay you to do it.* To be honest with you though, I was not thinking of trying to influence how you make your choice or whether or not you go into research. It may be selfish, but I am looking for ways to tell you about my peculiar type of creative problem solving. People like my dad and I add our creative products to the pile of progress, but no one learns very much about or from our process because it is hard to explain and we are not very verbal. That is my impression, anyway.

MFJ: I can tell that you want to get on the record some of your many quirks or weaknesses so that when you start hitting me with knacks and strengths, you will not feel as guilty about bragging.

MW: Again, I say you are perceptive. My mother's two main rules for me were: *Be humble;* and *Let us not talk about it anymore.* Even as I say that, I realize that it comes over as colder than was really the case. The word 'humble' comes from the word for 'ground'. From my father, I may have taken some lessons about how to use discontinuity. From my mother, I learned the more critical lesson that the continuity of a good life depends on how well you stay grounded.

BREAK

MW: Well, John, here we go again. This is the second time we have recorded in the wee hours of the morning. I am glad you saw the light on back here and came on over. I would apologize for the hour, but if you saw the light and chose to come, it means you weren't sleeping.

MFJ: No. I think my sleep is as irregular as your own. If you want to know the truth, I have been a little worried about you. You seem to be agitated and want to get everything out as fast as you can, but the associations keep piling up faster than you can get them out. Maybe we need to talk about our process to see if I can take any of that pressure off.

MW: Just knowing you are sensitive to this helps. Things have gone in ways I had not anticipated. Let me see what I can suggest because I am losing sleep and aware of being agitated or irritated for no reason. I feel like I am exposing my inner thoughts and I fear that to you or to that mythical reader, I am coming over as a silly child or an old fool.

I believe in using what the unconscious provides, so let me tell you the memories that popped into my head.

- I once did a fire-walk workshop in which the last thing was to walk on a bed of hot coals. As it came my time to walk, I was not fighting physical fear as much as I was fighting social fear. I imagined a headline in the paper that read, *Idiot Psychologist Gets Burned.*

- I was living in the country with my grandfather during first grade. By the time I attended second grade, my mother had moved (with us three children) to the small city of Greenville, SC. All of my classmates had learned to write cursive in the first grade: in my first-grade class, we had only learned to print. Because of not being able to write cursive and probably because of other ways that I was backward or shy, someone said I was an *ignoramus* and that was shortened at some point to *Igin*. Throughout grade school, that was my nickname. Mr. Bishop, my best friend's dad, always called me *Igin*, though he may not have known how I got that nickname. When I went to high school, I got away from the nickname and do not want it to catch up with me again. At some point in life I learned the mantra, *I am just ignorant about that; I am not stupid.*

Maybe I should stop this monologue and give you a chance to get in a few words.

MFJ: No. My job right now is to listen and I am content in that role. My guess is that it takes some pressure off to just get some of these things out. Some of them may also help me to understand you as a researcher. What we are doing here is research, too, and I see your determination to go wherever it leads.

MW: I am really getting to be an old man. You say something kind to me and I get tears in my eyes. You are a good friend. Polly has also been very supportive of these sessions with you. She says it has put the light back into my eyes. She has been most helpful by reminding me that I can say what comes to me now and trust the editing process to correct, add, delete as needed.

She, in fact, edited a monograph on archeological sites for the National Park Service (Worthy 1983). I trust her judgment. She has also said that when typing the transcripts of these sessions, she learns new things about me or about eye color. She never knew or had forgotten that eye color is related to shyness until she read it in our last session. That is a group difference between blue-eyed and brown-eyed children and does not say anything about me or about any other individual.

You have to be careful not to assume anything about yourself because of some group statistic. Blacks, for whatever reason, have not dominated golf as they have some other sports, but thank goodness Tiger Woods did not look at the statistics and decide he should go into basketball. This week, as you probably know, he won his fourth Master's tournament and he has done so at the youngest age that that has ever been done.

Speaking of our last session, my dad jumped into the conversation unexpectedly and, for once, his presence made me feel *more* secure, not *less*. I found myself identifying the two of us as a pair and that has taken a long time. I remember when our daughter, the first of our two children, had her first birthday. It could have been after one month or one year, I am not sure. I was in graduate school in Gainesville, Florida. On the way home from the university, I stopped to get her a card that would be just from me. Outside the store I used the trunk of the car to write on. When I signed my name, *Daddy*, probably for the first time, I became physically ill. I had to wait a while before I felt in control enough to go on home for the celebration. I spent the waiting time thinking about the kind of daddy I wanted to be. Whatever we get accomplished by looking at eye color or research, if I am starting this late in life, to accept my dad at a deeper level, that is progress.

MFJ: Like they say, we strive for progress, not perfection.

MW: Oh yes, one day at a time. As *they* say, of course. Not that we have any problems that serious ourselves. Wait until I am in your role. I will know where to aim my questions.

MFJ: You seem more relaxed. More like yourself.

MW: I feel that way. I feel tired, but it is a good kind of tired. I hope our tennis drills are not rained out tonight. I need the exercise and that has become an important part of my social life.

You said during our last session, when I mentioned my father's invention, you would look it up on the internet. I *Googled* it last night. I was wrong when I told you it is called the Worthy Filter. The correct name is the *Worthy Strainer*. The manufacturing company still carries the Worthy name, though the company has changed hands several times since it was founded by my dad and his brother-in-law in 1951. My dad's idea, his *aha* insight, his puzzle solved, was that you could design a strainer to fit on a spray gun and strain "while you paint." That solution to a messy, time-consuming problem has stood the test of time. There was information about the Worthy Strainer in catalogues and in *Wood Digest's Finishing* magazine issue of June 2004. A Mr. Roger H. Phelps of Phelps Finishing in Bristol, Wisconsin is quoted there as saying, "I've used

Worthy Strainers (made by Worthy Mfg.) for over 20 years. I wouldn't spray a finish without one on my gun." [A more complete history of the Worthy Strainer appears in an article, "Taylor Taylor makes worthy Worthy Strainers" by Michael Dresder, that was published in the Woodworker's Journal eZine of April19-May 2, 2005, Issue 122.]

Last night, I kept thinking about all those products in all those catalogues that represented in each case some person's chance to make a small mark on the total record of progress. He or she saw something that others had missed and, perhaps like my dad, despite all his weaknesses and failings, made sure his insight was not lost. I hope someone in the future can look back and say something like that about me. And, yes, I am talking about my *aha* experience of seeing that there is a meaningful eye color/behavior pattern in animals.

MFJ: I know that you made other creative contributions, which, if we both live so long as to get to them, will give me insight about your creative process. Why do you care so much about eye color of animals?

MW: Thirty years ago, I gave up every line of research I was doing then to concentrate on this one area. To me, there is a clear pattern in nature that is being missed or ignored. A 1973 article about my research in the *Science* section of *Newsweek* was the first mention in print of my hypothesis as it applied to human and animal behavior. My first book on the subject was published in 1974. The human aspect of the reactivity hypothesis received validation from the seven studies done at Penn State. So far, there has not been the same level of independent research to test the animal part of my hypothesis. Without independent verification, my ideas are only ideas. I got my hopes up some years ago when a top biologist at Oxford University e-mailed me and said that his graduate students were on holiday, but when they got back, he planned to ask if any one of them wanted to study eye color as a variable. He had the type of background in multivariate statistical techniques applied to animals at different taxonomic levels to give the hypothesis a rigorous independent test. I have not heard any more from him, but sooner or later I expect that he or some other person with similar credentials and resources will check it out.

I want to go on right now and state that when I use the word "expect." I mean it only in terms of "anticipate" and not at all in terms of anyone having an obligation to test my ideas. I once had a minor disagreement with another psychologist because I said something about what I "expected" in the first sense and he thought I was saying something about obligation. Only after he chided me did I realize that once again, I was using a word that might be correct in some sense, but was still wrong because I had failed to recognize that what was being heard was different from what I intended to say. That has happened to me so often that I think it illustrates one of my quirks. Let me go on to say that, being socially or perceptually slightly out of kilter, can also be a useful tool in giving you a unique insight, but the costs are real.

MFJ: That reminds me of something you said earlier about quirks of language. Can you give me another example? I see exactly what you mean with the word "expect", but I would tend to think of that as a single incident rather than necessarily evidence of some trait or quirk you have.

MW: Let me tell you about two episodes in my life in which I used the words *open* and *free* as completely synonymous and embarrassed myself. The first one happened at the tennis court next to the church. I had to have been about 12 or 13-years-old. I know that because by then my father had left for good and was living in Chicago. For my 12th birthday, he sent me a Wilson tennis racquet. It was the last time he acknowledged my birthday in any way. Only now when I say that--and surprise myself by starting to tear up--do I make the connection to the emotional response I had when I signed my daughter's first birthday card.

John, again, I feel like I am getting into more than I signed on for. I keep exposing my feelings. Does that make you uncomfortable?

MFJ: Not at all. We agreed that we would let these conversations go wherever they choose. We both seem to have less control over the process than we anticipated. Your willingness to get into areas that are emotional for you may give me the courage to do the same when the time comes one day for you to interview me.

MW: Thanks. That lets me know I do not have to feel like I need to protect you. I trust Polly and she says doing this is good for me.

MFJ: I will play my role and say we had a thread going about some *quirk* of how you got embarrassed by using *open* and *free* like you had *expect* that was not normative and led to some kind of social problem.

MW: Thanks again. A number of boys and girls my age hung out at the church tennis court—across from the Sans Souci Baptist Church--the "without care" Baptist Church. Given our age, one activity was learning to court--at the tennis court. We were having our first semi-pseudo-quasi dates. The church youth group would schedule parties from time to time, such as a churned ice cream social at the tennis court. A date for us would be to arrange with a girl to walk with her to the party and then walk her home when it was over.

One day, I decided I was ready to ask a girl for a date to such a party. My pick of who to ask was a girl named Mary Jo. She was friendly, attractive and, most important of all, lived just a few doors away from the church. Somewhere, perhaps in a book, I had learned that one way to ask for a date was to ask the girl if she was free on the day you had in mind. That sounded like the polite way to start the conversation. Unfortunately, as it turned out, other teens were within hearing range when I got up my courage and gave it my best shot. I walked up to her and asked, "Mary Jo, are you open Saturday night?" I don't think I have to tell you how much teasing I got. And poor Mary Jo had to blush her way through it all and let me know, kindly I must say, that she could find her way to the party without me as an escort. I did learn how to ask, though, and had my share of dates, but not with Mary Jo.

Years later, after I was married, I made a similar mistake by confusing the same two words. I and several brothers-in-law decided to stop at a bowling alley to see if we could get in a game or two. We could not remember if this was a night for only league bowling or if some lanes would be available for "open bowling." Since I was the first one to reach the rental desk, I asked the man there, "Is there free bowling tonight?" Even after I tried to explain my mistake, the man was not very understanding. He let us know how much money it takes to build and run a bowling alley. My brothers-in-law enjoyed his lecture more than I did. Bullets:

- Even as I was relating that to you, John, I started to wonder if the more acceptable term is bowling lanes rather than bowling alley. Maybe "bowling alley", as a term, is dated, regional, determined by social class or some such.

- That expression, "some such," was a favorite of my Papa Morgan, the part Native-American grandparent, lifetime farmer, who warned me that I was getting too much schooling for my own good.

- Terms like "some such" capture how I feel about the eye color and behavior pattern with animals. I may not be doing a good job of describing it, but I am confident that I am pointing in the right direction.

- And that reminds me of the Zen monk who said that when he pointed out a shooting star to his students, they never saw it because they kept looking at his dirty fingernail. All I have known to do for 30 years is just to keep on pointing, to keep on gathering more archival evidence and, now with this, I am taking a shot at pointing to the pattern one more time.

Is the term "bowling alley" still used?

MFJ: Who cares, really? Are you a good bowler?

MW: Terrible. One year, my wife and I bowled on a team in a mixed doubles league. She was invited back for the next season. Football was my game and I still have the bad knee to prove it.

I am beginning to feel that I am presenting myself as more inept than I really am. The language problems, though, are real. I almost did not go to graduate school because my spelling is so spotty. I can know how to spell a particular word one day and then not be able to spell it the next. What saved me was that I just happened to read that Andrew Jackson was a poor speller and said, "It's a damn poor mind that can think of only one way to spell a word." That was my ready quote then and it gave me the courage to go to graduate school. I have always tried to have some last-ditch defense in reserve.

MFJ: I am going to remember that Andrew Jackson line for when I need it. I know already from things you have told me that you believe there is an up side to your wayward way with words. You used it to write puzzles and in hearing metaphorical language when doing psychotherapy.

MW: That is absolutely true. My skills or knacks are the flip side of my quirks. I miss some things that are obvious, but sometimes I notice things that perhaps others have missed.

MFJ: Can you give me some examples?

MW: I will give you two examples that illustrate the point. One deals with words and one does not.

I was reading in the book *The User Illusion* about some split-brain research that had been done at Cornell University with a sixteen-year-old boy, P.S., who prior to surgery, had suffered from severe epilepsy. I read these words:

"P.S.'s right and left hemispheres did not always agree. For example, his left brain would announce (through speech) that P.S. wanted to be a draftsman when he grew up, while his right brain spelled its way to 'racing driver" (Norretranders, 1998 p 279).

That example was used to illustrate how the part of the brain that controls conscious spoken expression may not have all the answers in the total brain about what is important to us. The point, as stated by the author, is that the two answers do not agree. When I read the two answers, I had to laugh because to me the two answers did agree. What jumped out was that *draftsman* and *racing driver* are synonyms. To be good as a racing driver, you have to be good at a type of drafting that differs from what one usually associates with *draftsman*. I have no idea if the people who did that research picked up on that or if it is more than coincidence.

However, it is similar to how unconscious information can be present in the conscious mind in disguised form.

Now you will tell me that that is a good example alright—of a psychologist with a warped mind.

MFJ: You misjudge me. Yes, I may make fun of you, but I am serious about understanding creativity. I want to learn all I can about what you consider your knacks.

MW: Well, for my second example of a knack, I want to tell you about something that happened to me in college. Looking back on it, it may have been the first time I began to realize that I might have a knack for sometimes seeing patterns that others have missed. I was taking a final exam in biology. The exam consisted of a long list of multiple choice questions, perhaps 80 in all. I answered the ones I was sure of and left the others blank to study more carefully. As I started to look back over the test, I got the feeling that there was something strange about the test. I saw that it had nothing to do with the questions as such. It involved the sequence of letters I had written as answers. They looked familiar to me, for some reason. Then I had my *aha* experience. The professor, for ease of scoring, had repeated for the second half of the test the same sequence of random letters for correct answers, a, b, c, or d, that he had used in the first half. The correct answer for question number 41 was the same as question number one, forty-two and two, forty-three and three, etc. Knowing that, I was able to fill in the places I had left blank. When I turned in the exam, I told the professor that I thought I had gotten all the answers correct and how I had done it. He was not as amused as I thought he would be.

I have thought about that incident many times. It captures how I get ideas or solutions, but it is not easy to explain. I sometimes pay attention to a strange feeling or know somehow that there is something to see before I see it. That is as close as I can come to an explanation. I think of it as my unconscious mind alerting me that it, with its resources, has seen something of interest that I have missed.

If you really care about learning how the unconscious mind works, be sure to find a copy of that book, *The User Illusion*, originally written in Danish by Tor Norretranders in 1991. There is an English translation (Norretranders 1998). It is based on solid research. I was impressed with just how much we now know about the unconscious mind and how counter-intuitive so much of it is. Read it. There will be a quiz.

MFJ: Yes sir. All kidding aside, I notice that your voice changes when you start talking about someone else's work. You do not hedge as much and you speak with more authority. Your voice seems to change a lot depending on whether you are talking about yourself or someone else.

MW: You are hitting close to home and I find I am very tired. We, or I, have to get soon to the topic of those gun accidents I mentioned. Since I brought them up and know we will talk about them, I cannot get the memories out of my mind and it is disturbing my sleep.

MFJ: Now is a good time to stop, but just tell me this before we do: Is there anything else I should know before we get to that topic? About my role, safeguards, or anything like that?

MW: John, that is a wise, caring question. I need to reassure you that I have professional resources available to me if I need them. I would stop these sessions before I put myself in danger. I also trust my unconscious mind to look out for me. I am touched though by your asking me the question.

There is just one other thing I need to talk about before I talk about guns. That is a quirk I used to have until about 14 or 15 years ago. It has to do with my right thumb. Trying to understand that quirk led me somehow back in time and a realization of how episodes with guns influenced my life.

MFJ: Your right thumb. Okay, I think I can remember that. Until next time. Get some rest. It may be good timing, too, that we are playing round robin tennis tonight at Fair Oaks, just to relax, have some fun and look at the women. Right?

MW: Watch it there. You know Polly transcribes these tapes. The next time I say I am going to one of these mixed doubles events, I will have trouble getting a note from home. See you tonight.

❖4

MFJ: I hope all that tennis helped you sleep.

MW: I slept solid until 4 a.m. and awoke ready to get this over with. Then when I was ready to come in here and call you, I had to wait a few minutes and started going through my old personal files. I kept finding things that were relevant to our sessions, like pictures, letters, and the like and I kept pulling them out. The next thing I knew, two hours had passed. I wonder if it was just another way for me to stall. The last few days I have thought about almost nothing else except the things we need to talk about. My hair stands on end, I get goose bumps. I am about ready to start talking about my thumb and then guns. I just want to tell every episode and be done. Then we can get back to eye color and my other research. I hope you have a lot of time.

MFJ: I have plenty of time, day or night, for the next week. You can go at whatever pace you want and I will adjust to it. You said you were "about" ready to start. I suggest we go on to the

kitchen now, get a bite to eat, then bring hot coffee back with us. We both function better with coffee and once we get started we might not want to take a break. What was that T.S. Eliot line again?

MW: Good plan. "The evenings, mornings, afternoons, I have measured out my life with coffee spoons." As he says, talking about the "butt-ends of my days and ways."

LATER

MFJ: Now that I have my coffee, I am ready for the duration. For this part of what we are doing, I think I need to more or less stay out of the way and if you need me to be more involved or less, just tell me. You were going to tell me about your days and ways, starting with something or other about your right thumb.

MW: What I want to do is tell you about many episodes. Trying to understand the episodes about the thumb led me to see at a deeper level the role guns have played in my life, even though I have never owned a gun other than an air rifle, or as we called it, a BB gun.

For each episode, there is a story, which I will try to tell as short and to the point as possible. I think of getting these stories out like bullets in a written presentation. I may jump from one to the next without any words of transition.

About 30 years ago, in the mid 1970s, our family spent a week's vacation at a beach house on the coast of South Carolina. As I was loading the car and getting ready to come home, I asked the children to make sure that the family dog, Trixie, did not get out of the house. She might run away and we needed to get on the road back to Atlanta. Someone forgot, she came running out, I lost my temper, took a dive to grab her, missed her, landed on my thumb and once again, went away with an aching right thumb. I had hurt it many times before. My only broken bone, ever, was my right thumb. My only night spent in a hospital was because of my right thumb. I had hurt that thumb so many times before, I decided then and there to work on understanding why. Since we were starting on a long drive and my wife was willing to do most of the driving, I had a good opportunity. Rather than read, I just kept my focus on the hurting thumb to see what images or memories would come to me. I had become sold on that technique during an earlier family vacation spent on a boat in the Bahamas. I guess I should tell that first.

Early one morning on that trip, I went up on deck where several people had chosen to sleep. The bald head of a man was just visible inside his sleeping bag, which was green on the outside and red on the inside. There was something about that scene that gave me a very strange feeling. I walked around to the other side of the boat and sat down. I sat alone for maybe an hour just looking out into the early morning mist. I just held on to that strange feeling without trying to make sense of it. After some time, I had an image or memory of something I had seen during the year that I had lived with my grandfather and attended first grade in 1942/1943. The image was of a postcard that my uncle in the army had sent to his dad, my grandfather. The card, which sat on the old family pump organ for a time, showed people eating watermelons. One young woman is noticeably pregnant and is shown saying, "I must have swallowed a seed." Then, sitting on the boat, I remembered something I had totally forgotten. At one time, as a small child, I had been told that babies came from watermelons, and believed it. In fact, as a small child I was the only one in the extended family who did not like to eat watermelon. It was clear to me that the scene with the man's head coming out of the sleeping bag triggered a feeling because it activated something inside from that time in my life.

How a green and red sleeping bag led to a memory involving watermelons became one of my favorite stories to illustrate how the unconscious mind works. However, there was another part of the story I avoided telling until now. Later that morning, after the sun was up, some of the strange feeling was still with me. To see if any other images or memories came to me, I stretched out on the now warm deck. I lay on my back with a straw hat over my face and looked with half-closed eyes up toward the sky. The hat had plastic "eyes" or openings for ventilation. The light from the sun made each opening a small bright circle. As I relaxed and went into a twilight zone, every one of those little bright spots suddenly turned into a skull. Those images were so clear and persistent, that it was very disturbing. I removed the hat, got up and got back into the routine of the trip. It is clear now that going back in memory to that time in my life tapped into more than a theory of babies and watermelons.

Back in the car riding home from the beach, I stayed with the feeling in my aching thumb and thought about one of the earliest thumb injuries. Toward the end of my last year in grade school at Sans Souci School, the principal, Mr. Bennet, planned a special event for only the seventh grade boys. The only part of that day I remember was a three-round boxing match in which I represented my class and Buddy Garren represented the other seventh grade class. I thought I would lose, but I was prepared to give it my best shot. It went much better for me than I expected. For whatever reason, I was more than holding my own. In the middle of the second round, the feeling of dominance or whatever, gave me a rush of energy and perhaps arrogance. I threw my best right, hit him with a glancing blow, and sprained my right thumb. I did not let on, but the next time I hit him with my right hand, I knew it hurt me more than it did him.

I quickly decided that I would only use my right to fake and hit with my left hand. That worked fairly well for the rest of that round. In the third round, he was all offense and I tried to hold him off with left jabs. I was determined to make it through the round without quitting or being knocked out. Buddy got in a lot of good shots, but the round finally ended and I was still standing. Mr. Bennett, acting as judge, said I won the first round, Buddy won the last round and the second round was a tie. He called the match a draw.

During my sophomore year at North Greenville, then a two-year college in Tigerville, SC, I was captain and quarterback of one of the intramural flag football teams. Many of us had played high school football and took the game seriously. Our last game of the season was against a team on which I had many friends. We had no chance to win the league title, but if they beat us, they might end in a tie for first place. The day before the game, one of the players on the other team came to me and suggested that we should make sure they won the game since it meant more to them. I said, "No way." Nothing arouses my righteous indignation more than poor sportsmanship. I did not tell anyone else about that conversation, but I was even more determined than before to beat them if we could. The game was close; we were driving down the field, when somehow I managed to break my right thumb. I was taken to a hospital and came back, a cast on my hand, to learn that we had lost the game.

When I was in the Air Force, in Germany, one of the doctors had me stay overnight at the hospital for observation because of a red streak going up my thumb toward my hand. Whatever the cause of the red streak was, it was gone the next day and I was released.

It is time for a break.

MFJ: You were telling me about times you hurt your right thumb.

MW: There were many other times I hurt that thumb, but I no longer remember the circumstances. I have no idea what the one in

Germany was about. This may be crazy, but I just heard the word "about" as "a bout". That is the kind of metaphorical language I learned I had a knack for hearing when I was working as a counselor. The example from seventh grade dealt specifically with a bout. Maybe there was some type conflict involved in Germany that I no longer remember.

Riding in the car from the beach, I realized that often hurting the thumb had served to stop me from exercising strong dominance in a situation in which I was feeling very aggressive. Even the episode with the dog getting loose had triggered a sense of outrage unusual for me to express toward our children and the hurt thumb stopped that emotion. It was as though hurting my thumb stopped an emotion that might hurt someone in some way.

Then, riding in the car, or perhaps later, I remembered something that had happened one day when my mother was home from work. I am not sure how old I was, maybe eight or nine. I was outside that day playing with other children when my mother suddenly called me in to the house and scolded me. She had been watching from a window. She complained that I had let a younger boy push me around. She sent me outside and said she would be watching; I had better not let it happen again. The first thing I did when I got outside was push the biggest boy there, two years older than I was, into a mud puddle. He came out swinging, as I knew he would. I fought back with fury. He finally won, as I also knew he would, but by then, I knew that my mother had seen enough to stop her concern that I was a coward. Looking back, I think I was more afraid of hurting someone than I was of getting hurt myself.

Riding in the car back from the beach, I focused on the emotion that had been stopped that morning, the one that stopped my prize fighting career with Buddy Garren, and the one that stopped my attempt in flag football to be the avenging angel that ended the other team's chance for the title. I sensed that in all three cases, there was an aggressive emotion which hurting my hand had stopped. In its strongest form, it might be called *fury*. As we neared home, I just let myself be with that feeling and I kept coming back in my imagery to a scene when I was five years old, living in Pelzer, SC. Focusing on that emotion put me back there in my front yard, lording it over the little boy next door.

By the time I was back home unloading the car, I knew I had to sort out the events of that day and how it changed my life. The best way for me to do that is to start with that day in 1941 and then, tell stories, bullet-style, most of which involve guns, until a type of closure of sorts in 1991 and then, the writer's block that started in 1996 or 1997 that you have helped me to unfreeze. I wanted to set that up before we took our break so that when we come back I will know where to start.

John, thank you. Just your presence allowed me to talk. Say something, even if it's off the wall. You had better take the chance because when we come back I want to spit out as many butt-ends of my days and ways as I can in the time we have left today. Speak, Sir, that we may record your voice as proof that you were here.

MFJ: I do have something and it is off the wall. I remember last night at the tennis center social, someone brought pieces of fruit and I saw you eating watermelon. Right?

MW: Right, you are. I must have finally come up with some other theory about the origin of babies. However, just in case, I am still careful not to swallow a seed. Let's take a break.

BREAK

MW: Well, I have switched to ice tea and I am ready to talk.

MFJ: I have switched to a diet coke and I am ready to listen.

MW: The day was Tuesday, December 2, 1941. I was five years old. The time was between noon and 1p.m. That, I learned later. I want to stick to just what I remember. I came out into our front yard and saw a pint milk bottle that someone had thrown into the shallow ditch that separated our small front yard from the street. I also saw that the little boy next door, Tommy Pearson, was in his front yard. I picked up the milk bottle and said to Tommy something like, "This is not our trash. It must be your trash," and threw the bottle into their front yard. Tommy said it was not and threw it back into my yard. We kept throwing that milk bottle back and forth. I felt good. I was going to win this battle. Tommy must have been getting more and more frustrated because he said, "I will just get a gun and shoot you." He went into the house and when he came back, he had what I thought was a toy gun. I was standing at the edge of our yard. He came over to where I was, pointed the pistol at me and said, "Now I am going to kill you." He tried to pull the trigger. Nothing happened.

He moved back toward his house as he continued to manipulate the pistol. I stood in the same spot between our two houses watching him. Suddenly there was an explosion that I will never forget or entirely get over. The bullet went into his face and up through his head. To say that I saw an explosion is the right way to say it. I must have stood looking no more than a split second. My next memory is of running between the two houses and into the back door of my house. I could not find my mother. (She had been inside working at her sewing machine; she had heard the shot and went out the front door to check on me.)

I remember only one other thing. I looked out the back window, or back side window, and saw men coming toward the scene, running on a path that ran to the street behind us. One was my father; he was wearing high top brogans. It is the only time I can ever remember seeing him or a group of workmen run like that. It was terrifying. I have no more memories of that day.

When my mother could not find me outside, she went back in our house and found me sitting on the floor in a back room playing. She assumed that I had come back into the house when Tommy had gone back into his house. I did not tell her or anyone else that I had been there when it happened. It was my secret. My world changed on that day.

Everyone's world changed five days later when the Japanese bombed Pearl Harbor and the country went to war. My mother's two brothers were already in the military. One was on the aircraft carrier USS Lexington, when it was sunk in May 1942; he survived.

Those uncles fought in the Pacific. My father and his two brothers also soon went into the military and fought in the European theater.

Before that, though, in early 1942, my father went on one of his alcoholic binges that had the effect of us again having to leave a town, this time Pelzer. Because my mother was sick and needed to care for my little sister, my older brother and I went to live with my grandfather Morgan and his wife, my step-grandmother, who lived on a farm in Pickens County-not too many miles from Greenville where my mother was living. That arrangement was only for the 1942/43 school year; I was in the first grade.

One day when my mother was at the farm, she took my brother and me with her to visit neighbors, the Long family, who lived on a nearby farm. My brother and I went with one of the sons about our age, Henry Long, to his bedroom. He showed us his 410 shot gun. He assured us it was unloaded and started pretending he was hunting. As he swung the gun around, he said, "Yonder goes a rabbit," and pulled the trigger. Again, I saw and heard an explosion. This time it was only a wall that suffered the damage. Before his mother or our mother could get there my brother and I were out the front door and ran all the way home.

There are many other things that happened that year on the farm and most of them are very good memories. The only other thing I need to mention, though, is that all the stress did take a toll on me. It probably helped to determine the fact that I was so shy. I would often not speak up, even when it was in my interest to do so. I also had a classic hysterical symptom that was probably related to my having a secret. I was always trying to clear my throat. My mother was convinced that there was something stuck in my throat, but the doctors said there wasn't. Finally, she took me to a chiropractor, Dr. Georgiana McDaniel, who saw me for a number of sessions. I have no memory of those sessions, but the throat-clearing symptom ceased. When I went to visit her 37 years later and told her that I had decided she was a psychotherapist as well as a chiropractor, she laughed and said she had been accused of that many times. Whatever she did worked; I finished first grade. My father was now in the army. My mother was now well and my brother and I moved back to live with her and my sister in Greenville.

The third gun accident involved an "unloaded" air rifle, which my cousin aimed at my face and pulled the trigger. The BB came through a windowpane and just missed my left eye. He was more upset than I was. We managed to cover the hole in the window such that it would not be noticed by my aunt and uncle. At least this time I had someone who shared in the guilty secret.

After the war, many things happened when my father came home and started his alcoholic binges again. After a year or two, he left for good. There are many bad memories from those years. I will only tell one—a gun story. He had been gone for some weeks. I was about 11 years old. I was at home alone one day when he showed up drunk. I stayed in the house with the doors locked as he pro-ceeded to the detached garage, which he had used earlier as an automobile paint shop. There was still a lot of equipment there, including some he had built himself. He took an ax and smashed up the whole place. When he started toward the house, I was not sure what he had planned next. I had come up with a plan, though. I took the magazine out of my air rifle so that the barrel would be empty. I opened the back door a crack and stuck just the end of the barrel out. I yelled, "I have a shotgun and if you don't leave I am going to shoot you." He may have been too drunk to see clearly, but for whatever reason, he left. My big fear had been that he would rush the door and I would not have time enough to pull the barrel back in and re-lock the door.

Over the years, I had many nightmares about trying to get that door closed as someone or something struggled to get in. It was always something dangerous at the door-often a dangerous animal. One I had to laugh about when I woke up. It was a cobra in striking mode. The most prominent feature was the cobra's wide *hood*. That was funny to me because my father's name was Hood Worthy.

John, my man, thank you for hanging in there with me. When we meet again I will pick this up to show where it led. From now on I will be talking mostly about how I dealt with the residuals of these episodes and how I managed to get some closure. I feel a great relief from just telling you about the childhood events. When the gun memories intruded on what you and I were trying to do, I started to feel a great weight. I feel like much of that weight is now off.

MFJ: This has been some day. Thank you for letting me share it with you. You said the gun memories intruded on what we were trying to do. They fit very well with what I am trying to do.

Could you have possibly found a better way to illustrate the researcher's feeling of being intrigued with the phenomenon. My guess is that the way you tackled the thumb phenomenon that led you to all this other is a true picture of how you work. I remember you told me that you did not, by training or initial interest, set out to study animal behavior, but were led by things you found to drop other research and concentrate there. To me, there are recurring threads in all we have talked about.

MW: Good enough. I am starting to like those threads of yours. I will also get back to the thumb thread. The connection of that to the gun experience may be obvious to you by now. We may even get back to more of a conversation and not so much a monologue. At least I know enough not to make any promises. These conversations will go where they want to go. It seems like you and I are just along for the ride. Good night.

BREAK

MW: Here we are again. If you don't mind, I will pick up the gun theme again in a minute. First, why don't you ask me some questions--any questions--so that we get in some dialogue before we go back to my gun stories. Dialogue seems more alive to me.

MFJ: Do you think that being alive is more important to you because of knowing that what happened to Tommy that day could easily have happened to you. The gun could have gone off when it was pointed at you. Did the close call make life more precious to you?

MW: That is logical and I would like to say yes, but it would not be the truth. It is not how I experienced it. My take on it is, a dangerous world seemed even more dangerous and part of the danger came from me. If I was not careful I could cause someone to die. That is heavy response ability (two words) for a five-year-old to bear. At some level, I think I believed that I was the one who should have been shot. Think about it. At that age, punishment often hinges on answering the question, "Who started it?" I knew, I know, and I will always know that I started it. I made it happen. Adult logic puts it in a whole different context with multiple causes. Knowing all that helped cover up the emotions planted at age five, but it did not remove them. Think about the sleeping bag/watermelon story; it comes from the same age. The strange feeling or emotion I had on the boat that day followed from associations and beliefs I had given up consciously and forgotten decades before. On the surface, my 15-year-quest was to understand how "the accident" in Pelzer had affected me. At a deeper level, it was to understand how "the killing" had affected me.

MFJ: "Who started it?" I think you are right; that is at the core of morality for five-year-olds.

MW: To me, hearing you state the question just now, "Who started it?," brought back the memory of looking out the window at the men, including my father, coming on the run. The terror I felt about that scene must have been from fear of what would happen if they learned the truth about who started it. Somehow, I must have decided not to tell and did not for more than 30 years. It worked as a defense against terror, but there was a cost. Some part of my life was lost that day also. F. Scott Fitzgerald (1945) describes coming through an emotional crisis as it having all worked out, but it worked out for a different person. The book, *Zen and the Art of Motorcycle Maintenance*, written by Robert Pirsig (1974), has as a major theme the author's trip back to Montana to learn or to remember who he had been before he became mentally ill and was cured by shock therapy. He had found sanity, but he had found it as a different person. It was something like that I had in mind when you and I talked in an earlier session about my having lost my voice, or some such thing.

MFJ: My guess is that literature has been important to you. Are there other writers that come to mind?

MW: There is always William James. His 1899 essay, *What Makes a Life Significant* (James 1925), tells how, "Every Jack sees in his own particular Jill charms and perfections to the enchantment of which we stolid onlookers are stone cold." He uses that simple fact to make the point that there is much to know of value in people around us if we just open our eyes. There is a line in that same section of the essay that asks, "Where would any of us be were there no one willing to know us as we really are?" And here you sit with your coffee mug willing to listen to me "spit out the butt-ends of my days and ways." Shall I pick up the thread?

MFJ: Do that.

MW: Until 1973, I would tense up any time I was near a pistol. At Georgia State University, where I worked from 1966 through the end of 1993, I could observe on a regular basis how I responded to being near a pistol. To get between two parts of the campus, the easiest way was to use the crosswalk on Courtland Avenue. Most of the time there was a campus policeman there to stop traffic. I noticed that each time I came near him and his holstered pistol my body would become tense for a few seconds. Something happened on July 4th, 1973, that ended or greatly weakened that response.

That day, we took the children to a nearby resort, Lake Lanier Islands, for a picnic. Because of the large crowd, one area was already closed to cars when we got there. There was a chain across the road with a metal sign attached announcing that this road was closed to all but foot traffic. We decided to unload our car there. The family could stay with our picnic gear while I found someplace to park and then walked back. As I unloaded the car I had not really paid any attention to the fact that my six-year-old son was amusing himself by shaking the heavy chain across the road, which rattled the metal sign. The first thing I became aware of was a security officer with holstered gun running down a hill toward us, screaming angrily. I thought he was angry at me for parking there and something, maybe his running toward me and his gun, caused me to momentarily freeze. Then, as he got closer, I realized that I was not the cause of his wrath, but my son was for shaking the chain and rattling the sign. That released a fury in me. I went out to meet the security officer, screaming at the top of my voice. I asked him if he had no more sense or training than to come, armed and running, to scream at a child. I berated him fairly and unfairly. I asked him if he really believed my six-year-old son was going to break his iron chain. He must have thought he had encountered a crazy man. He finally managed to get in the words, "When you are finished, please move your car," and left. Only later, when I one day noticed that my body no longer became tense when I was in the crosswalk near the policeman's pistol, did I realize that somehow that Fourth of July marked the start for me, too, of gaining a measure of freedom or independence. That brings tears to my eyes.

MFJ: I feel more ready to jump in with my comments now that we are talking about adult experiences. When you were telling about your childhood, I felt, to some degree, you were back there and the best thing I could do was just to be here as a quiet, supportive adult. With this, I feel like we can talk more man-to-man. I would say that the protective emotions of you as a father stampeded your emotions left over from Pelzer. The fact that the man was running toward you seems significant. That, and the gun. What is really interesting to me is that it had a more general effect of weakening your fear of pistols in other situations. Does that fit?

MW: It does. And I will give another response to what you said. It illustrates how vigilance gets primed to detect certain stimuli. When, a moment ago, you used the word *childhood*, I heard what you were saying and, also, the word *hood* jumped out. You remember that was my dad's name.

MFJ: I am intrigued with what you have said about metaphorical language and how important things in the unconscious get aired in disguised ways. Think about your favorite expression on the tennis court.

MW: You mean *40-love* when I am serving?

MFJ: No. If you or anyone else hits a really outstanding shot you will say—

MW: "One shot, one life." That is a Zen saying related to archery. If you keep practicing and get out of your own way, from time to time, you will make a shot, in archery or tennis, which makes it all worthwhile. Playing the game is just the excuse you need to be out there where such a magic moment can happen. One shot, one life. With all that, I think you are right that the expression is over-determined for me, and that may be why I use it so much. It captures what happened at Pelzer that day. It was, indeed, one shot, one life.

I just had another insight that I want to share before it gets away. I just realized that earlier in our sessions I talked about three gun accidents as a defensive way to lesson the emotional power of the one that mattered. Almost as if I wanted to add the other two to lower the average impact of childhood gun accidents in my life. If that is so, that is weird. Of course, not all such "*ahas*" are correct.

MFJ: It might be better for you to resume the story before we get too bogged down in *maybes*.

MW: It was a few years after the picnic episode that I hurt my thumb at the beach and riding home realized that the thumb episodes were related to feelings going back to the Pelzer experience. I started the uncovering process that was more or less completed 15 years later in 1991. I both wanted to uncover these things and at the same time, had a lot of resistance to doing so. I had to face some things that hurt to know or remember again.

One thing I did was to ask my brother-in-law, Robert Huff, who worked at the Greenville News, to see if he could find anything about the shooting in the archives of the newspaper. He found a news article and an obituary. One item that surprised me was that they knew Tommy had been playing with a "little friend" when he went into the house to get the gun. Nothing was said to indicate whether the friend was still there when the accident occurred. Did anyone see me start the argument about the milk bottle? Did anyone see me run away? Those were concerns from the five-year-old frame of reference. Reading that reference to his "little friend" made me feel vulnerable. That made no sense from an adult perspective.

Something else hit me much harder. The newspaper said he was four years old. Having started to talk some about the Pelzer event, I always said of Tommy that he was a "little boy my age who lived next door." Because I knew I had been five--three months short of six years old at the time--thinking of him as younger than I was put an even worse light on it for me. Prior to that, I had thought of it as a fair fight that I had started. If they were right about the age difference, I had to have played the role of bully and been even guiltier. I could only hope that the newspaper was wrong. Years later, I learned his true age and it turned out that they had been wrong. He was actually only three years old; had he lived, he would have had his fourth birthday two months later in February of 1942. I had faced the possibility that he might have been four; now, I had to face the fact that he had been three. From the frame of reference of a five-year-old, that is a huge difference in age. I had to really work on getting to the adult perspective of seeing us both as little children doing what little children do.

Only after getting there to some degree could I go on. Some years went by and, then, I faced another issue that I had forgotten, although I have already told you about it as a fact. For some reason, I feel like this may be a good stopping place for us today. I felt driven to get to here. *Here* is having told you, confessed to you as it were, that Tommy was only three years old. If you hear that and can still look at me with soft eyes and listen to the butt-ends of my days and ways, I know we can make it together through the rest of these sessions. Some hours of rest may be useful in knowing what path to take. Right now I am not sure.

MFJ: I like that idea of using intuition to know when we need a long break rather than be dictated to by the clock. Was there something else? It was like you started to say something earlier, just as I started talking.

MW: There is one more story that may fit in here somewhere. I will tell it and be done with it. In my mind, what happened in Pelzer happened because I was acting high and mighty (one of my mother's favorite terms). Earlier today it occurred to me why, or perhaps why, I picked up on that T.S. Eliot line about "butt-ends" of my days and ways. In 1945, my dad came home from the war and he and my Uncle Roy started a business painting cars in the detached garage. My uncle liked to tease me and having been a sergeant in the army, he could also cut me down to size in a hurry if I tried to get too smart with him. I was nine years old and one day, I had what I thought was a bright idea. I had found a smoking pipe somewhere or other. As my dad and uncle both smoked cigarettes, there were butts in ashtrays in the house. I tore some of them apart and filled my pipe, got some matches and went outside to a good hiding place beside the house. I lit up and was very proud that I had introduced myself to the manly art of smoking. My uncle, who had fought in the Pacific islands, must have known more about good hiding places than I did. The first thing I knew, he was standing over me with the garden hose. Before I could get away, he thoroughly soaked me, and my pipe. John, if I start sounding too high and mighty with all the butt-ends of my days and ways, I am depending on you or Polly to get the garden hose.

MFJ: Count on it.

❖ 5

M FJ: Good morning. Where do we go from here?

MW: The only thing I know for sure is that I want to start off very slowly reviewing the process before I get caught up in telling more stories. Yesterday, I spent most of the day in some conflict about whether to go on with this project; if so, what direction to take and whether to continue to think of it as something that might be published.

MFJ: I will tell you now that I hope we continue, but it is entirely up to you.

MW: I can imagine some potential reader recalling the Shakespeare line, "Me thinks he doth protest too much." That is part of what makes this difficult—my being intimidated by some potential reader who could laugh at me or look down on me in a way that he could not if I just kept my mouth shut. With you, I see the soft eyes and I want to tell my stories and, truth be told, I want others to read my stories. And yet, when I am away from here, I start to think about all the times I misjudged the reception I would get when I told something or other about myself, my research or my experience in life, that I start to feel defensive and want to hedge in every way I can. Unless the reader is as interested in the psycho-logical process as I am, he is shouting about now, "For God's sake, get on with it. Who do you think you are, Kafka?"

MFJ: You once told me that you like to hike along ridges on the Appalachian Trail and be able to look down on either side. You said to stay sane (that's the word you used, *sane*), you try to always stay on the ridge between feeling like a "sad sack" on the one hand and a person with delusions of grandeur on the other. It seems you fear a reader will judge you one way or the other. If a reader did not like to read about psychological process, would he or she not long ago have stopped reading this and gone on to something else. This may be one of the areas where you are still locked into thinking like a five-year-old child— that you have this power to hold an audience which will all the time be condemning you. Cheer up and assume that any readers that you get beyond Polly and me, and maybe your grandchildren, are just gravy. So there, take that. I think it irritated me that you considered bailing out on me. I have an investment here, too, you know. *I* am interested in process, if no one else is. I have spent hours willingly doing this because I want to. I am excited about what I am getting out of this personally and maybe professionally, and you would have thrown me aside because of fear of being criticized by some

unknown reader whose library has only this book on the shelf. As that lawyer, Sullivan, said at a congressional hearing, "I am not here to serve as a potted plant." Surely, you are not so much into your own stories that you have failed to realize that I am using this as a model of how to get into some things *I* want to talk about when the roles are reversed. There is more to me than the confident self I show on the tennis court. Do you not know that?

MW: Well, I do now. I guess I was comparing my insides to your outside. People make that mistake with me, too. Be assured that when the roles are reversed, I will be thinking about you and not myself—just as you have done for me. I will be listening, then, really listening. I like your emotional fire, by the way, and welcome it. Do look out for yourself.

MFJ: Something happened to you over the weekend. Would it help me to understand you for you to tell me about it? I am aware that my job is to probe, but it is also to respect your privacy. Is there something there to tell or should I go on to some other topic?

MW: Tomorrow, I may tell you about the process I went through Sunday. Today, I want to just give you some observations or conclusions I reached. I am aware that I need, as part of our process, the practice, from time to time, of just doing what I call bullets; and, yes, I am aware now of the metaphor. I need to get things out without thinking too much about how they fit. I know we have agreed already to do that sometimes. The problem I get into is being overwhelmed with so many associations going off in different directions and feeling some need to make this about eye color. That was my goal, and I remember that Jack Wright warned me about being like a hound dog that cannot stick to the same track; that dog never gets to chase a tired rabbit. The solution I came up with was this. Because I like this dialogue process, I can think of doing it for as many years or books as I last. I will bring this first effort to some kind of closure and then decide whether or not to publish it. If you go back to graduate school or get tired of doing this, I may try my hand at what Walker Percy did and just interview myself.

When we started, Polly noticed immediately that this was cathartic for me. That is reason enough for me to continue. I was totally blind to the fact that this had great value to you. You were right to set me straight. I know now you are not just doing charity work for an old man.

MFJ: Far from it. If I may be assertive, let me suggest that you stay with the technique of free association. Just jump around all you please. There may or may not be readers for that, but I get to hear your stories and what is being alive anyway other than telling each other our stories.

MW: You sound like Jack Wright. That was his philosophy. He said that my puzzles would appeal to people who are crazy in the same ways I am crazy. Perhaps the same could be said for my stories, as well. Others, too, live with traumas and memories of doing such things as asking their own Mary Jo if she is going to be open Saturday night.

MFJ: I hope you will do some of your "bullets" or free associations now. The idea fits with everything you have said about yourself: you think by *ahas*, you cotton to discontinuity in thinking, jumping rather than smooth transitions. Just be yourself. Just run up your own flag and do not worry about who does or does not salute.

MW: That I can do. I will just see where my free associations take me. I enjoy seeing Eleanor Clift on TV news programs. It reminds me that she interviewed me about my eye color work with humans and animals for an article that appeared in the *Science* section of the November 19, 1973 issue of *Newsweek*. That article was the first thing ever published about my eye color research. I want to read you the last paragraph of that article. It contains my hypothesis and some of my philosophy of life.

The burden of his research, concludes Worthy in a book to be published in January, is that "dark-eyed organisms— human or nonhuman—specialize in tasks that require sensitivity, speed and reactive response, while light-eyed organisms tend to specialize in tasks that require delay, hesitation and self-paced response." But the 37-year-old psychologist is careful to note that his conclusions apply to groups rather than to single individuals, and that between self-pacing and reactive skills, neither is more desirable than the other. "Dark eyes are not superior to light eyes nor is the converse of that true," he cautions. "Neither can reactivity nor non-reactivity be considered as superior in any absolute sense. If we are to truly appreciate and respect talents different from our own, we must recognize the degree to which we are enriched by individual and group diversity."

Eleanor Clift also reviewed my 1974 book for the June 3, 1974, issue of the newsletter, *Race Relations Reporter*. The review ended in this way:

> Refusing to do instant personality analysis, Worthy bristles at even the hint of such treatment of his work. "It's meaningless to judge an individual on the basis of eye color just as it is to judge on the basis of race," he argues, "but we can still talk about group averages."

> A low-key, deliberate person in conversation, Morgan Worthy is thorough and careful almost to a fault. He explains himself backwards and forwards in academic language that does not lend itself to snappy quotes. The blue-eyed Worthy is definitely self-paced. And he probably knows more about eye color than anybody else.

> Fearful that his work might "undermine feelings of brotherhood and equal worth," Worthy addresses its social implications in his book's introduction. While he dreams of America's pluralism being enlarged to accept and respect

genetic differences in behavior, he half-jokingly suggests that
the three graduate students who assisted him (one blue-
eyed, one brown, one hazel) draft essays on "the natural superiority of light eyes," "the natural superiority of dark eyes,"
and "in all things moderation or the best of both worlds."

I could not have asked for a reporter to be fairer in presenting
my hypothesis and its social implications. Notice that I was talking
about the value of diversity in 1973. However, eliminating discrimination should not be assumed to result in some predictable
uniformity. Even back then I recognized that if we were to truly
accept diversity we had to be prepared to respect cultural *and*
genetic pluralism. That is consistent with current scientific research. Read, for instance, Matt Ridley's 2003 book, *The Agile
Gene: How Nature Turns on Nurture*. Nature and nurture influence
each other in such complex ways that it is folly to assume some
single cause of observed group differences. Scholars need to feel
free and safe to offer possible interpretations that may later be
proved wrong. As Ms. Clift wrote, I dream of such a world. By the
way, John, her brother-in-law was the famous movie star, Montgomery Clift.

**MFJ: Wait a minute. That is strange. Let me try to play your
role of reporting an association.**

MW: Lord, help us.

**MFJ: The first Montgomery Clift movie that popped into my
head was *Suddenly Last Summer*. The theme of the movie was**

about a woman, played by Elizabeth Taylor, who had to confront a very traumatic memory. Montgomery Clift played the young psychiatrist. Is it possible that even your name-dropping is over-determined or influenced by your unconscious mind?

MW: That is impressive and I do remember that movie, but I cannot even start to give you an answer.

MFJ: Do you not recognize a rhetorical question when you hear it? Go on with your history.

MW: My primary training in scientific research was in social psychology. In 1969, I and two of my undergraduate students had published in the *Journal of Personality and Social Psychology* a laboratory study of exchanges of self-disclosure (Worthy, Gary and Kahn, 1969). Because the article was later determined to be one of the most cited articles in social psychology, I was invited to write a *Citation Classic* article (Worthy 1987) and used the opportunity to honor two of my graduate professors, Sidney Jourard and Jack Wright. From Jourard I got the hypothesis and from Wright, I learned the laboratory technique.

That was the last laboratory study I did because also in 1969, I read the book, *Unobtrusive Measures* (Webb et al. 1966). It made a case for some folks getting out of the laboratory and start doing studies with data already available in formal and informal archives. I knew, at once, that that was for me. It more closely matched what I had done and loved doing in military intelligence.

In 1970, I gave up teaching and took a position in the GSU Counseling Center. I also published with one of my graduate students, Alan Markle, an article (Worthy and Markle 1970) based on archived sports statistics. We suggested that there was a pattern to black/white differences in college and professional sports, with blacks doing relatively better on athletic tasks that required quick reactions and whites doing relatively better on athletic tasks that did not require quick reactions, but rather could be done at one's own pace. Later, I came to the conclusion that those observations were better understood in terms of differences in eye-darkness rather than differences in race, as such.

One thing that is very important for every one of us to remember is that *for understanding nature as a whole, eye-darkness will always be a more useful variable than is race.* Eye-darkness has similar meaning across species; racial categories are limited to one species.

MFJ: I see what you mean. Scientists value most highly findings that generalize most widely.

MW: I could not have said it better myself.

MFJ: I am going to treat that as a deserved compliment and move on. You had gotten as far as 1970.

MW: By 1971, I had discovered, through more archival studies, some differences between dark-eyed and light-eyed whites in sports performance that led to the reactivity hypothesis later confirmed in the studies of reaction time done at Penn State (Hale, et al., 1980). I might have stayed focused on that human area of research for the rest of my career, but I found that the same pattern is present when comparing dark-eyed and light-eyed animals on such things as hunting tactics and escape tactics. Exploring that vast area became the most important part of my research.

Switching my interest to animal behavior took me so far outside my own primary area of training that I qualified for amateur status. I think I felt more at home there. The ideas of amateurs do (for good reason) tend to get ignored, but I was free to play with data the way I had learned in the Air Force and from the book, *Unobtrusive Measures.* Equally important to me personally, because of fall-out from the nature/nurture conflict, I had lost some trust in the peer-review process. This was, after all, the 1970s.

This is my final bullet or free association of the day. During our long break this afternoon, I watched an episode, on *BBC America,* of the comedy, *Keeping Up Appearances.* I like it because I can identify with all the characters. In this episode, Hyacinth and her husband, Richard, are going to an up-scale auction. She tells Richard, "Today may be the day when someone mistakes me for someone important. If someone mistakes me for aristocracy, don't tell them any better. It would be bad manners."

See you tomorrow.

BREAK

MFJ: Maybe we should plan our afternoon breaks to coincide with the daily episode of *Keeping Up Appearances* on *BBC America.* You really enjoyed telling me about that yesterday.

No doubt what you remembered reflects some deep truth about your inner psychic workings.

MW: Either that or I like good comedy programs on TV.

MFJ: Do we go back to the early 1970s where we left off, finish with the guns theme or pick a new track and chase a fresh rabbit.

MW: Later today, or maybe tomorrow, I want to get back to what's left to tell about guns. You may have guessed that I had to get off it for a day or two because some images were coming back too strongly when I tried to sleep. After getting onto another topic and then, taking Sunday off to rest and play, the images faded and Sunday night I slept for eleven hours. After we tie up some loose ends from things I got into yesterday, we can go back to the gun stories.

MFJ: I hope the loose ends you have in mind include why you became disillusioned with peer-reviewed human research in psychology.

MW: No one sees the absolute necessity of peer-review more than I do. It makes sense that no finding is considered a part of the body of scientific knowledge until there is independent evaluation and verification. For a person who works alone as I have, confirmation bias is always a potential threat. That is why I value so highly the studies of reaction time at Penn State that provided strong evidence with humans that the reactivity hypothesis points to a real phenomenon. Those studies were entirely independent of me.

During the years of my career, the most rigorously peer-reviewed journals in psychology were those sponsored by the *American Psychological Association*. If you wanted to shoot higher, you could submit a manuscript to *Science* in the United States or *Nature* in Great Britain.

By 1970, I was proud of the fact that I had broken the ice by publishing two articles in the *APA* journal for social psychology, *Journal of Personality and Social Psychology*. However, there were several things that I observed which let me know that the profession was changing.

I was changing, as well. In 1970, I decided that my shyness and anxiety in front of groups was never going to go away. By that time, I had taught a year of high school, taught undergraduate courses part time at the University of Florida and William and Mary College in Virginia, and then, taught undergraduate and graduate courses full-time for four years at Georgia State University. When a position became available in the Counseling Center at Georgia State, I applied for it and was accepted. It required much more of my time, but it was mostly working with individuals—work which I loved. After that move, there was less demand that I publish and I was doing more of my research after work hours. I could afford, and, in fact, needed to do research that was fun and exciting to me. I needed a release from the pressures of the clinical work.

MFJ: So you were changing and that coincided with your perception that the profession was changing also.

MW: Yes. And I want to be careful to state that this was my opinion based on several things that I observed rather than coming from some inside knowledge or systematic study. Interpretations based on nature rather than nurture appeared to receive strong resistance in the social sciences.

MFJ: So, you felt like there was some double standard operating when it came to advancing nature vs. nurture interpretations?

MW: It was an indication, to me, that we were in danger of correcting an extreme in one direction by going to the opposite extreme. It seemed that now nature was out of favor and some people in the profession wanted to force all human behavior into cultural interpretations. Scientific findings were evaluated, in part, I believe, by the degree to which they agreed with that position. I observed several relevant things about the same time; I am not sure of the chronological order.

At a meeting of the *Southeastern Psychological Association,* a discussion was chaired by a professor who was very active in publishing review articles of current research on group differences. He asked for feedback on something that he was considering, but had not fully made up his mind. The idea for discussion was that people involved in the peer-review process had an obligation to see that nothing got into the literature that might undermine *higher societal values.* A heated debate followed. I had to give him credit for alerting us to what might be coming. At the same time, it seemed contrary to everything I had been taught about science and peer-review.

When I found some archival evidence of behavioral differences among whites related to differences in eye color, I thought that it was a finding I wanted to get into the peer-reviewed literature. The problem was that the findings would seem to increase the possibility that some of the observed group differences in sports behavior might be genetic in origin. Would the results conflict with higher societal values? If the paper was rejected, would I really know why?

MFJ: I see the problem. Unless the peer-review process is transparent, you will not know how it is that a rejected manuscript failed to come up to standard. You could work to improve the quality, but you would be wasting your time.

MW: Exactly. And, also, subtle or unstated types of discrimination are the most harmful, psychologically, because they kill your motivation and make you paranoid. I got into an argument with a journal editor about that time and felt like he was applying a double standard. That is hard to judge fairly, though, when you have a personal investment. I chose to walk away from the argument because of what it was doing to me--making me more suspicious than I wanted to be.

MFJ: So what did you do with the eye color manuscript?

MW: I decided to shoot high and submitted it as a possible report in *Science*, the journal of the *American Association for the Advancement of Science*. They rejected it, which was not that much of a surprise, but what they said did surprise me. I pondered their words. Shortly after that, Robert Helmreich, a professor from the Psychology Department at the University of Texas, was visiting Georgia State. He said to me, "Morgan, the manuscript you sent to *Science* created a storm of controversy at Texas." He told me that the manuscript had been sent to a former president of the American Psychological Association for review and he got together a committee to advise him on whether or not to recommend that it be published.

I said, "Bob, just answer one question for me, please. The rejection letter had a line about having finally decided not to publish the report *at this time in the nation's history*. Did that part about *this time in the nation's history* figure in the final decision?" He said, "That was the deciding factor."

MFJ: It is clear that you feel that appealing to *this time in the nation's history* is not scientific. Were there other things that made you move away from the peer-review process?

MW: For one thing, by moving to animal behavior, I put myself outside the area in which I could claim to be a peer, based on training. That was made clear to me, but that is a story for another day.

There was another experience I had in that same time frame that I should tell you about. It cut deeper in a way and made me try to think of my research as something I did for my own enjoyment. I told you of how influenced I had been by the book, *Unobtrusive*

Measures, which taught a research philosophy akin to what I had been taught in the military. I identified with that philosophy. The authors were all at Northwestern University. One, Lee Sechrest, I got to know from his visits to Georgia State and from seeing him at meetings. One day he said to me, "I have noticed an interesting phenomenon. When I am invited to visit places to lecture about our book, *Unobtrusive Measures,* I always mention your research as what we had in mind. People get very interested until they ask, 'Where is this guy?' When I answer, 'Down in Georgia', they lose all interest."

MFJ: There is not much you can do about that.

MW: No. I understand it. I have some of the same bias. Someone sent me a clipping from a book review in the *Boston Globe* in which my puzzle book was described as "highly entertaining and reward-ing." Then, I got a letter from the editor-in-chief of *Atlantic Monthly* saying that his family had enjoyed working the puzzles in my book during a long car trip. Some of the pride I felt in getting those responses had to do with geography.

John, I still have not said what it was about that comment from Lee Sechrest that affected me so. I was disappointed of course. In addition to that, I realized that it provided me with one more excuse to explain away any failure I experienced. You can learn to use things like that to feel helpless. Learned helplessness can lead straight to depression. You have to believe that the game you are playing—research like any other type game—is based on a reality that allows you some chance to win. It does not have to be a fair chance or a great chance, but enough of a chance that you do not feel helpless.

The point I want to make is that I cannot afford to focus on some possible bias and take my eye off those ways I have failed to do a good job in presenting my ideas. I cannot afford a good excuse. Before you accuse me of attempting false modesty, let me remind you we are talking about helplessness and depression. As long as I know I am the main cause of any failures, I am not helpless. I am thus not depressed. If I felt differently, I would have to find another interest in order to avoid feeling helpless and depressed.

Let's go hit some tennis. What do you say?

MFJ: I say I will not take the hint. If I beat you so badly in tennis, as usual, that you feel helpless and get depressed, that is your problem, not mine.

❖ 6

MFJ: Here we go again. I feel like there are so many loose ends, I am not sure where to start. I also know that you want to get back to the aftermath of your gun experiences. Which should we do first?

MW: Ask me your question and I will decide to start with that or come back to it later.

MFJ: You said that you adopted the research philosophy presented in the book *Unobtrusive Measures,* because it was akin to what you had been taught in military intelligence. Could you give me some more details about that philosophy? I want to get straight what that means to you.

MW: The idea is that initial steps in a new area of research should learn from geologists and first look at "outcroppings" of easily obtained, or already available information before setting up expensive studies to get new, precise, information. Let me read to you from their book. "Any one such outcropping is equivocal, and all types available should be checked. The more remote or independent such checks, the more confirmatory their agreement," (Webb et al.1966, p.28). Their idea was so out of favor in social science that the working title of their book was *Oddball Research*.

MFJ: With that title, no wonder you felt like they were talking to you. It is worth thinking about though and it makes sense to me. Another useful thing I am learning from you in this process is how you look out for yourself or protect yourself psychologically. I had always thought of defense mechanisms in a negative sense, as a kind of blindness. You seem to consciously set up defense mechanisms. Like your rationalization, "It's a damn poor mind that can only think of one way to spell a word."

MW: Yes, the Andrew Jackson quote that gave me the courage to go to graduate school. Actually, I had a second quote in reserve, in case I needed it. According to a biography I had read, Leonardo di Vinci was noted for being "individualistic" in his spelling. If one excuse did not work to reframe my poor spelling, then perhaps the other one would. Later on, I learned from Georgiana McDaniel an all-purpose, tertiary, line of last defense: *If I didn't do the best I could, at least I did the best I did.* That one has helped to keep me sane.

MFJ: Sane?

MW: Yes. I am willing to be thought of as *crazy* in a figurative sense; I just do not want it to be certified—at least, not until I can claim senility. In some way, that saying from Dr. McDaniel helps me to get unstuck from the past and move on.

MFJ: It is always a surprise to me what we get into. I think we are going to talk about one thing and we end up talking about something else. That is not a complaint at all. That is part of what I like about doing this. Do you think it is possible that one reason you started talking about defense mechanisms is that we are about to get back to the traumatic experiences you had as a child and unconsciously you wanted to be reminded of your defenses, or however one would express that?

MW: You may be right. The unconscious mind works in mysterious ways to get its work done. When I got to graduate school and met my classmates, there were two guys who unintentionally intimidated me because they were so verbal. They had used their undergraduate time well and could recall more about psychology than I could recognize. Drinking coffee with those guys, I remember saying when they asked about my goals for graduate school, "I plan to stay in the middle of the herd and hope that none of the professors learn my name."

Now, note that that was a conscious defense, but no one is aware of his unconscious defenses. Let me read you a sentence from the preface to the book, *User Illusion,* "In recent years, scientific investigations into the phenomenon of consciousness have demonstrated that people experience far more than their consciousness perceives; that they interact far more with the world and with each other than their consciousness thinks they do; that the control of actions that consciousness feels it exercises is an illusion" (Norretranders 1998 p. ix).

You are right. I start in one direction and end up somewhere else. Maybe I am just stalling.

MFJ: Anything else before we finish off the gun saga.

MW: Last thing. I promise. I like stories about coming from nowhere or coming from behind to make your mark. I think of Silky Sullivan, the racehorse that always came from behind to win races, or Billy Mills, the Native American runner who broke out of the pack to win an Olympic gold medal. Watch it on YouTube. Earning graduate degrees felt something like that to me.

MFJ: I think the racehorse you are thinking of is Seabiscuit.

MW: It may be true of Seabiscuit, too, but the horse I was thinking of was Silky Sullivan, a California horse that won a lot of races, but lost the Kentucky Derby. That loss probably has a lesson for us, too. Enjoy your wins because you never know which one will be your last.

MFJ: One shot, one life.

MW: Exactly. When I earned my Ph.D., it felt so strange I did something quickly before the university could have second thoughts. I drove to Columbia, SC to see my dad at the auto body shop where he worked. He and I enjoyed our tour of the shop with him introducing me to his co-workers as, "My son, Dr. Morgan Worthy."

Silky Sullivan, indeed. Alas, to show off for a while at the auto body shop does not mean you will win the Kentucky Derby. No matter. One shot, one life.

Are you ready for us to pick up the gun, so to speak?

MFJ: I am. You had told me about the gun accident that happened in Pelzer when you were five years old, that you tend to think of as "the shooting" or "the killing" because you felt responsible. Then, there were other gun experiences you told me about. We had reached a place where you were about 40 years old in the 1970s and had found a news account indicating that the boy who died was younger than you rather than the same age as you. You remembered the event very well or some of it very graphically, but you had repressed that he was, in fact, only three years old. That was tough for you. You felt guilty for having started the argument that led to the accident.

MW: Yes. These things we will talk about next will cover roughly 15 years, ending in 1991. I will tell about them and not worry too much about whether or not I get them in the right order. It took so long because I was busy and also because it was painful to look back at that event. That was especially true after I had to face the age difference between the two of us. It made it even more significant that I had started it.

Then one day something else dawned on me. Not only had I not told my parents at the time about my having been there when the gun went off—I still had not told them. After I had been grown for a while, I told some people about the Pelzer event, but I had never talked about it with my mother or father. That was the next thing I had to face.

First, I told my mother. Her first response was exactly what I knew it would be. She had heard the shot and immediately went to the front door and out to look for me. I was not out there. I told her how I had run between the two houses and in the back door. Once she realized that as she had gone out the front door, I had come in the back door, it all fit together and she accepted it. She had more trouble accepting the idea that my digging up all this pain from the past was a good idea. I was not sure, either, but I knew I was going to do it, even if part of me did resist and delay the quest.

I next went to see my dad to tell him about it. I found him in a V.A. hospital suffering from emphysema. He was very interested in hearing all about it. When I told him of looking out the window and seeing him and the other men coming on a run on the path, he remembered it well. He said that he had been working at the Ford dealership, which had been a few blocks from our house. I did not tell him how terrifying it had been for me to see those men coming on the run. At least, I had now told my parents.

Finding and telling his parents was another matter indeed. There is something about that I have to tell you. Telling my parents was relatively easy; when I told each of them, it was an adult talking about his childhood. When I thought about telling his parents, it was difficult to stay in an adult frame of mind. It was difficult to stay in the third person, as if looking at it through my eyes now. I would start seeing the images I saw in person rather than seeing (third person) the child I had been. The fear and reluctance I felt was that of a five-year-old child rather than that of a 40-year-old man. The next time I was ready to renew the quest, I did not go looking for his parents--I went looking for the chiropractor who had treated me when I was in the first grade.

John, I think that needs to be it for today. Just thinking about these things is draining. I want to have time to get a nap and eat before it is time to go to tennis drills.

MFJ: Be sure to get that nap. Once we get on the court, I do not want any excuses.

BREAK

MFJ: Here we are in the wee hours of the morning again. How is it going?

MW: Same old, same old. I start talking about these memories and my sleep gets disrupted again. I keep remembering things and I am determined to keep to the thread and not tell you unnecessary stories. I woke up thinking about how sneaky or cunning my dad was. He would be drinking and call to say he would meet you somewhere and then when you got there, you could not find him. He would be hiding out of sight so he could watch to make sure you were alone and had not brought the police.

Maybe I will tell you two stories. They are the only two stories that my parents, even my mother, could tell and laugh about. The first has to do with how they left Pelzer. My father had been arrested for public intoxication and was placed in the Pelzer jail. He made such a fuss yelling, in the jail, that my mother was mortified. By the next morning, she had already rented an apartment in

Greenville and had hired a van to load and move the furniture. She planned to be gone before he got out of jail. As it happened, he got out earlier than she expected and hid near the house. As the van pulled away, he sneaked up unseen and climbed in the back. When my mother and the workers went to the back of the van to unload, my dad was there, in the truck, sitting in an easy chair.

This is the problem, John: I start telling a funny story and it reminds me of a story that is not funny at all. My friend, Charles Sizemore, who lived one street away, also had an alcoholic dad who was sneaky. His father came home from a binge and was angry that no one was there. What he did not know was that his wife was in a hospital; in his mind, she had stayed out all night. When she came home the next day, he jumped out of a closet and killed her with an ice pick. I remember that and it reminds me again of the time I told you about when Daddy walked around the house all evening with a butcher knife in his hand. There are too many associations to too many things. One night when he was angry like that he knocked out all the windows. My mother was probably right that you just have to reach a point where you do not talk about those things and get on with your life.

So let us get on with it so we can stop talking about it. I do want to finish telling you about the 15-year quest to understand how the gun accident had affected me and what I did about it. I am glad I stayed with it then and I will stay with telling you about it now.

MFJ: You were going to tell me about going to see the chiro-practor who treated you when you were in the first grade and

suffering from a throat-clearing habit you later decided was related to your secret about the gun accident.

MW: Her name was Georgiana McDaniel. She was retired and living in Liberty, SC. I told her my name and told her she had treated me 37 years earlier when I was in the first grade. She asked, "Did you live with your grandfather out beyond Eighteen Mile Creek?" I had, and hoped she might be able to remember something about the case. If she did, she never let me know it. I told her that I had decided that in fact she had served as psychotherapist. She laughed and said she had been told that many times before. I liked talking to her and went back to see her several times over the years. I did not learn anything from her, though, about those early experiences. I did learn from her an all-purpose, last ditch, self-defensive statement. Quiz. Do you remember it? It began, "If I didn't..."

MFJ: "If I didn't do the best I could, at least I did the best I did." Next you will be saying that every time you lose a game in tennis.

MW: Why do you only mention my losing in tennis? I had my good moments last night and those are the only ones I plan to remember.

I am glad we got in that early morning time. Polly and I were tied up with our yearly eye exams today. I learned a new cliché that will come in handy as I get older. Our doctor said he asked a woman patient in her 80s how she managed to stay so happy all the time. She said, "Dr. Ross, you can be young only once, but you can be immature forever."

MFJ: Do you believe it really matters what cognitions you live by or what you say to yourself about yourself.

MW: Absolutely. There is an extensive literature on how we can talk ourselves into learned helplessness, which leads to depression. Brief advice: stay optimistic. It is healthier to believe you have a shot, even a long shot, in life than to ever let yourself believe you are helpless and have no shot of reaching your goals in life.

No more stalling. Back to *Guns* and *Hood*.

MFJ: Hood was your father's name, as I remember. You had a nightmare about a cobra trying to get in the back door of your house and realized a cobra symbolized danger and its hood symbolized your dad.

MW: I do not know that you can ever be that sure about a possible symbol, but that is what I thought and it allowed me to laugh about it, which is also important.

After my visits with Dr. McDaniel, the next thing that I remember that moved the story forward was in early November of 1989. I was 53 years old. One of my counseling clients mentioned wanting to be sure to talk with someone before he died. It hit me that I had some angry things I wanted to say to my dad before he died and he was already in his 80s. It had been maybe 10 years since I had talked to him. I knew he was living in Chester, SC with his wife of 37 years that I had never met. I called him that night or the next day and asked if I could come to Chester the coming weekend and take him to lunch. I said there were things I wanted to talk about. He said yes and we set a time and day for me to pick him up at his house. He mentioned that his wife had cancer, but I did not ask for details and the conversation was very short.

On the given day, as I drove from Atlanta to Chester, I thought about the things I wanted to say and ask. He had been sober now for 25 years, but I wanted to confront him about abandoning not only his wife, but also his children. As I drove, I started having flashbacks and my anger became mixed with fear. Then I remembered that the last time I had talked to him, he told me he had started carrying a pistol because he and his brother had had a big fight. I also remembered that the last time I took him to lunch years before, he had had me drive his car. I decided that he probably kept his pistol in the glove compartment of his car. I made two firm decisions. Having come this far and waited this long, I would tell him what I thought this day. Nothing would stop me. The other thing I decided was that I would insist that we go to lunch in my car.

When I got to his house, there were cars in the driveway and several men standing in the yard. My father's wife had died during the night. I asked why no one had called to let me know this was a bad time for me to come so I would postpone my trip. They said they had decided that it might be good for my father to see me and get away from the house for a while. Of course, they did not know the purpose of my trip. I met all his stepsons and stepdaughters and their spouses. Some of his grown grandchildren were there too, and they let me know how much they loved their grandfather. They had only known him sober.

When my father and I prepared to leave, he insisted as I expected that we go in his car. I remembered my concern about the pistol and insisted that we go in my car. Neither of us was prepared to give in. One of his daughters I had just met put her hand on my arm and said, "Morgan, he has to go in his car; that is where the oxygen tank is." I woke up as it were and started to see him as the frail 80-year-old with emphysema that he was rather than the bigger-than-life dangerous giant of my memories. I agreed that we would go in his car and told myself that if he made a sudden move toward the glove compartment, my hand would be on it before his. We left. He said he did not want to eat so we drove and found a place where we could park and talk.

You would think that a grown man would be able to see that conditions had changed and his resolution to confront his dad this very day should be abandoned. I was not in a frame of mind to see any value in abandonment. I poured out my anger and questions just as I had planned. He took it for a while and then his replies were as heated as my questions. I heard his side of the story. Not that I bought much of it, but it was a valuable experience for me. It was the rebellious adolescence I had never had—heated and totally irresponsible, considering that his wife had died during the night. I took him home and drove to Greenville. The next morning, my wife, my sister, her husband and I drove back to my father's house in Chester before attending the funeral that afternoon. My father decided he wanted the four of us to go with him to the mortuary that morning while the casket was still open. We did so and that is how my sister and I met our stepmother. Even the word sounds strange to me.

Several weeks later, I drove back to Chester to see my father. He had had a birthday and I took him a fancy barometer mounted on a nice piece of wood. My father did not seem thrilled to get it. I had tried to give it to my grandfather and he didn't seem too thrilled, either. (What do you give an 81-year-old man or a 97-year-old man?) No matter. My father and I spent the day in friendly conversation, man-to-man. That was a first and was very valuable to me. I think about conversations I have had with my own son and know how much my dad and I missed along the way.

I felt settled about myself, but as I thought about it, I knew I had some anger about my children missing the experience of having him as a grandfather in their lives. I both admired the relationship he had with his grandchildren and resented that my children had been deprived of that. Since my wife's father had died when she was 15, our children had grown up knowing two grandmothers, but no grandfathers. I decided not to confront my father with that since we had reached a point where we could talk and enjoy each other's company.

In December, we received a call that my father was in the hospital and my sister and I drove over to Chester to visit him. We did not know that it would be the last time we would see him alive. We had a very good visit until the last minutes we were there. As we were leaving and telling him to take care of himself, he said some-thing about some mildly selfish decision he had made and added, "I have decided it is time I started looking out for myself." Something about that flew all over me and before I could think I said, "That is all you have ever done—look out for only yourself." As my sister and I walked out of the room, he was lying in his hospital bed, glaring at me with hard eyes. I was glaring back at him with eyes that I am sure were equally hard. When we got outside, I apolo-gized to my sister for ruining what had been a good visit. She said she understood and we let it go at that.

By the next month, January 1990, I had thought a lot about the fact that I had confronted my father on the day that his wife died and again when he was seriously ill. I did not want to apologize or change anything I had said, but I did want to add something. I bought a thank-you card to send him. In it, I wrote that I wanted to thank him for hearing me out and also for the fact that he had been sober for 25 years. I said that that was a good example for us that, no matter how late, we could always change our life for the better.

Then, in February 1990, we got a call that he had died. Polly and I drove to Chester. One of his daughters gave me back the card I had sent and said he had been proud of it and showed it to whoever came in to see him. She also gave me back the barometer I had given him in November as a birthday present. That was the second time I had been given back that barometer after a funeral. With misgivings, I put it in the trunk of my car and my wife and I started the drive back to Greenville and then Atlanta. The highway went through the little community of Leeds in Chester County where the Worthy family had settled in the 1700s. Something felt not quite finished. I needed more closure. At Leeds, the highway crosses a river. Halfway across the river bridge, I stopped the car, assured Polly I had no intention to jump, and got out. I took the barometer from the trunk and threw it as far out into the river as I could. It floated and for a few seconds, I watched it bobbing down the river. Somehow that image helped because Polly and I were able to laugh about imagined reactions of fishermen as they saw that barometer heading downstream.

It is a relief to be done with that. Next time, I want to tell you about my return to Pelzer. I will be glad when we are done with all this and can get back to research and eye color.

MFJ: We have been talking about eyes. It sounded like you and your dad had to fight it out with words and then you had to have a final battle with glaring eyes that ended in a draw. I am struck by the fact that you almost waited too late to go back to confront him. You said you first went to see him in November and he died the following February. That is cutting it close.

MW: Indeed. I am one lucky guy. As they say, I have an *attitude of gratitude. They,* of course, are anonymous.

❖7

MW: As they say in Australia, G'day, John. Are you ready to go on with the Gun-shy Saga?

MFJ: Good morning. I know we must be getting close to the end of the 15-year quest, which you said ended in 1991 and we were up to 1990.

MW: Yes, I got some closure in 1991 and was more or less free of it until you and I started these recorded sessions and I realized that the gun-shy saga was what had caused an emotional writer's block in 1996 or 1997.

MFJ: You told me earlier that you pick up or drop terminology as needed. You have used the term, *gun-shy saga*, twice this morning; sounds like that term may be here to stay. You said that terms can be "over-determined" if they meet more than one need. I can see that "shy" may refer to being wary of guns and also refer to your shyness, which may have pre-dated your gun accidents. It serves to remind me that eye color is related to shyness. You may have been a "little boy blue" wary of

people even before guns started going off in your presence. How am I at free association?

MW: I am impressed. We will have no problem when we focus on you. That is, if you can free associate about yourself as easily as you do about me.

MFJ: Are you sure that what you do is free association rather than what the psychopathology textbooks call "cognitive slippage".

MW: As Jack Wright would say, "I will not diagnose you if you will not diagnose me." I resume the saga.

Perhaps dealing with my father freed enough energy for me to face going back to Pelzer. (Just saying that, I get a feeling like my hair is standing on end.) My mother remembered that they had lived on Park Street in West Pelzer. Knowing that, one day in 1990 (maybe November, but I am not sure), I drove to Pelzer and found Park Street. (I feel myself talking like a military report—make it as brief as possible and get the job done). I talked to neighbors to make sure I was remembering correctly which house had been the Pearson house. It was all as I remembered it. The Pearson house had suffered some fire damage and was no longer inhabited. No one was home at the house next door where we had lived. There was still a slight depression next to the street where I had found the pint milk bottle. I stood at the place where I had stood almost 49 years before and saw again in my mind's eye what I had seen that day. I walked to the back of the house—the route I had run that day. I saw that the path could be seen where I saw the men coming on the run.

Based on a tip from a neighbor, I located and talked to Tommy's aunt who had been nearby that day and she told me many details of what Tommy had done that morning. Tommy had gone with his mother and the aunt to the nearby cemetery to tend to the grave of his recently deceased grandmother. The aunt said that at the cemetery that day, Tommy had talked about being dead as any three-year old might and said he wanted to be buried near his grandmother. That was about more than I could handle. Either before or after talking to the aunt, I visited the cemetery and saw where Tommy was buried. Reading the age on the gravestone was the toughest part.

Later, as the 49th anniversary of the shooting approached, I was driving one day to Greenville with my wife and 26-year-old daughter. I detoured through Pelzer and showed them the places I had visited and we took some photographs.

It seems strange to use the word, *anniversary*, but, in fact, most years, I was very aware of December 2nd. That year, 1990, I was very aware of it. That morning, I got a call from a former client who had moved to another city. He was back in town for one day and had one free hour, from noon to 1:00pm. He asked if I could possibly see him at that time. It was the worst possible hour for me since it was during that exact hour that the accident had occurred 49 years before in Pelzer. I agreed to see him at noon and had no thought of mentioning what was on my mind. The hour was intended for me to listen to him—not tell him about my own concerns. I did not mention what was on my mind. As the hour went on, I was amazed at the turn the session took. He became very animated when he told me about watching a Doctor Bradshaw on television who indicated that adults could be affected by things that had happened to them as children. My client was fighting that idea. Perhaps he did not want to face something that had happened to him when he was a child. He asked me, "Do you really believe that we can still be affected by something that happened to us when we were five-years-old?" Those were his words and the age he gave—not four and not six, five-years-old. I did not want to answer the question because I felt he needed to answer it for himself when he was ready to face the answer. However, he would have none of it. When he asked me point blank for the third time, I said O.K., let me tell you what happened to me this very hour 49 years ago and how it affected me. Was it coincidence or synchronicity that he insisted on my answering that question on that day at that hour? I do not know. Did he get cheated of some of his time or was he well served? I do not know.

I do know that I had one more hurdle to face and that was even tougher than going back to Pelzer. Tommy's aunt had told me that Tommy's father was deceased, but that his mother was alive and lived near the North Carolina border in Marietta, SC. I had told the aunt that I had seen what happened that day, but I did not give her the details. I knew I needed to go see Tommy's mother.

MFJ: I have a suggestion. Rather than push on with the gun-shy saga, why don't you do some of your free association now and then, we can break to relax when it is time for *Keeping up Appearances*. Otherwise, we may miss the program or stop in the middle of your story. What do you say?

MW: Good idea. I can see, of course, that you are suggesting that in order to spread out some of the stress for me. I can also see that your idea is a good one. You are wiser than your years.

MFJ: Either that or I discovered a good teacher at the coffee shop that day. Give me an association; I want it free.

MW: I mentioned to you in passing about how important it is psychologically to stay optimistic and avoid getting into a rut of learned helplessness. That is one area where I think anyone can learn to live a more healthy life by reading relevant research. If you have any interest at all in that, track down the writings of Martin E. P. Seligman, who is or was at the University of Pennsylvania. He relates a lot of clever studies, some of them archival, some also with animals, which showed that how one perceives situations can lead to depression or giving up and that in turn may lead to an earlier death.

John, reading Seligman (1991), you can kill two birds with one stone. Not only can you learn valuable content, you will see a master at work applying the research approach I admire so much.

MFJ: Thanks. I try to give you a rest from the gun-shy saga and you end up talking about what leads to earlier death and killing two birds with one stone. The topic of killing will out, I guess.

MW: You are starting to listen with the third ear. I am impressed. I used to look for ways to teach that type of listening to counseling interns and practicum students I supervised. It is not easy to teach. Either you have a knack for it or this technique we are trying lends itself to your learning it. Then again, it may be a temporary flash in the pan, like your recent success in tennis.

MFJ: You wish. I am on to you. I even know why you identify so strongly with Hyacinth. No matter how others see her and how many times she falls on her face, down deep, she continues to believe that she really is someone important. With all the comedy, the program is about faith and coming from behind like Seabiscuit.

MW: To think I just praised your listening skills. Silky Sullivan. Do not switch examples on me. As for my identifying with Hyacinth, just remember that I said I identify with all the characters including the slob brother-in-law, Oslow, who sometimes just wants to have a beer and watch the telly.

It is time to turn off this recorder. That is all of this for me today. If weather permits, Polly and I are going to Pete's casino tennis round robin this evening at the Harrison Tennis Center. With my shoulder still hurting, I am working on hitting my serve at waist level or below and disguising speed and whether I am going to hit it with top spin or backspin. What do you think?

MFJ: You are addicted to spin; I agree with that. You might as well see if it works in tennis, also.

BREAK

MFJ: How was the tennis round-robin last night? I know Pete was trying this Las Vegas Style Round Robin where the winners of each match draw a card and at the end of the night prizes are given based on who has the best poker hand. How did that work out?

MW: Very well. In fact, there was an insight I had this morning about last night's tennis. And it may be relevant to what we are working on here, but it will have to wait. I want to get on through telling about 1991.

MFJ: You had gone back to Pelzer. You visited the place where it happened. You talked to his aunt about things he had done earlier that day and went to the place where he is buried. You had already told your mother and then your father that you were there when the accident occurred. You still had not contacted his mother and told her about being there when it happened. Did I leave out anything that is critical to the story?

MW: Yes. I had learned earlier from the old news story that he was younger than I was. On his gravestone, I learned that, in fact, he had only been three years old. That made everything much worse for me in my own mind. Whereas I had conveniently forgotten the age difference, it would have been something I would have been very aware of at the time. Now that I knew that he was that much younger, starting the bottle-throwing fight was less excusable. I had to face the fact that I had bullied or tormented a younger child to the point that he went to get a gun and shoot me. The accident would not have happened if the gun had not been available and loaded, and the accident would not have happened if I had not bullied him into a rage. It was great fun for me; I still remember the feeling of dominance; it was something else for him.

Knowing this, I kept putting off going to find his mother and tell her all about it. During this time, I was working long hours and had other responsibilities. I just kept putting off going to locate Mrs. Pearson. I was busy with work and family life, but it was more than that. There was some child-like fear I felt about facing her. Finally I gave myself a deadline. I would talk to her before December 2, 1991. That was the 50th anniversary of the shooting.

MFJ: You call it "the shooting."

MW: I know. There have been times when I thought of it as "the killing"; I am working toward being able to think of it as "the accident."

MFJ: Would it not be better and more accurate to think of it as *an accident* rather than *the accident?*

MW: Yes, it would. That would be an adult perspective that saw it as one of many experiences I had as a child. In 1991, it was difficult to hold on to that adult perspective. When I talked with colleagues, they would point out that it is not right to blame a child for acting his age, even if that child is yourself at an earlier time. That is easier to accept in theory than it is to apply in practice.

Finally, I had just about reached the fifty-year deadline and arranged to drive up to Marietta, SC on November 30th. Driving up I-85 that morning, it occurred to me that perhaps I should go by Pelzer one more time before driving on to Marietta. Since Pelzer is close to I-85, I would not lose much time, but I could not decide for sure. After I crossed the South Carolina state line, I decided I should stop at the next rest stop to get clear in my mind what I wanted to do before I reached the Pelzer exit. That rest stop has a path that goes beside a lake. I walked the path, decided to go by Pelzer before going to Marietta, and walked up to the parking lot. I nodded to a young man who was working on his automobile and he started to talk. I guess he thought anyone who took time for a walk on the lakeside trail could not be in much of a hurry. With no preamble or any comment from me, he said, "I had the strangest thing happen to me this morning. I live in Smyrna, Georgia, and got up early this morning to drive up to Pelzer to see my dad that I have not seen in five years. When I got on the interstate highway, I went the wrong way. I went toward Marietta [Georgia] when I should have been going the other way to get to Pelzer."

MFJ: What do you make of that?

MW: I report; you decide. I visited Park Street and the cemetery in Pelzer and then drove on to Marietta, South Carolina. I found the address with no problem and rang the doorbell several times. When no one came to the door, I turned to walk back to my car in the driveway. Between my car and me, there was a large Rottweiler dog. For a second, I thought I might have a real problem, but then I saw that he was on a chain and I could circle around him to get to my car.

I drove to a place where I could walk and debated with myself what to do now. On the one hand, I felt like I owed it to Mrs. Pearson to find her and tell her what had happened that day in Pelzer. Then I started making contrary arguments. Maybe I would be just opening up an old hurt that had had almost 50 years to heal. Since I had told her sister the year before that I had seen the accident, Mrs. Pearson could contact me if she wanted more information. Those arguments seemed genuine to me, but I also knew I was looking for any excuse to avoid facing her. After thinking about it, I decided on a plan. I would drive to Hendersonville, NC, and spend the night. The next day, I would drive back to Mrs. Pearson's house. If again she were not home, I would accept that it was perhaps better to drop it and not pursue the matter further. If she were at home, I would talk to her as I had planned.

MFJ: You decided to let fate take a hand and guide you?

MW: Maybe more like flipping a coin. Either way you think of it, I could not decide so I took the decision out of my own hands. The next morning, as I drove back from Hendersonville to Marietta, SC, I was listening to a radio program of swing music from the big band era. When one tune ended, the announcer said that what had just been played was recorded in 1941. At first, I thought that was a strange coincidence. Then I had to laugh at myself. This was, after all, a program of music from that era and that week was the 50th anniversary of the attack on Pearl Harbor. I reminded myself not to get too superstitious.

Mrs. Pearson was home, came to the door, heard my name and invited me in. Because her sister had talked to her about my visit to Pelzer the year before, she was not too surprised to see me. She said at once, "You know tomorrow is the 50th anniversary of Tommy's death." "Yes, Ma'am, I know that."

We had a long conversation and Mrs. Pearson was as under-standing and gracious as she could be. I was very glad she was home. There were two things about that conversation I want to relate. Both have been very valuable to me.

At some point, after I had told her about the events that led up to the accident, I commented that I had felt bad about it all those years, but I had tried to remember that I was only five years old and should not blame myself too much. She cut off my comment with a mother's response of "He was only three." Her response stung, but the tone of voice was just right. It put on me the blame due a little boy who had badly misbehaved. And as strange as it may seem, it took off of me the weight of feeling at some (five-year-old) level that I had caused Tommy's death and did not deserve to live myself. I had needlessly angered a younger child, but I was not responsible for his getting the gun or for a loaded gun having been left at a place that he could access. There had been plenty of regret and remorse that did not involve me. It was an accident and I was not the only one who looked back and knew, too late, that it could have been prevented.

The visit gave me a feeling of some closure. I say "some" closure because you can't erase those old programs, as I was to learn five years later, but you can drain them of some of their emotional power.

The other thing I learned was that the pistol had a stiff trigger. That is why the pistol did not fire when Tommy had it pointed at me and was trying to pull the trigger with his finger. It seemed evident that to pull the trigger, Tommy had used his thumb.

MFJ: That would explain why the barrel was pointed up toward his face. If you held a gun in the normal position and switched from pulling with a finger to pulling with your thumb, the barrel would tend to tilt upward. Do you think that is related somehow to all the times you hurt your thumb?

MW: I have it worked out in my mind, but there is no way for me to know if I am right or not. I feel reasonably sure that the two things are connected in some way. To start with, it was on my trip back from the beach when I really stayed with the feelings associated with the hurt thumb that I kept coming back to memories of that day in Pelzer. The feelings that I had prior to hurting my thumb each time reminded me (while riding back in the car) of the feelings I had before that first gun accident. At first, I thought I must have seen Tommy move his thumb to the trigger. Later, I thought of something else which seems more likely. I am sure I must have overheard my family or his family talking about the accident in the days after it happened. If they discussed his having probably done this with his thumb, I may have gotten some distorted childish theory of one's thumb being dangerous. Each time I found myself in a situation that elicited the same emotions I had had in Pelzer that day before the accident, it may have triggered an unconscious compulsion to disable my thumb before it could do any harm. That is a lot of "if" and "may." That is my best guess and I have not had any hurt thumb incidents since 1991.

There is one more thing to report about that trip to see Mrs. Pearson. When I got back to Atlanta, there was a message that I had had a call from one of my counseling clients. I returned the call. She said that she knew I studied birds and she had had a funny experience that weekend. A bird seemed to have adopted her. It had come to her window and would not leave. She gave me a description and asked if I could tell her what species it was. I agreed to look in my bird books and call her back, which I did. I read from an old book, *Birds of America*, the description of some particular species and she agreed that it fit. Because she wanted to look at a library copy of that book, she asked for the name of the author. I said, "Oh, the author is *Pearson* (1936)."

And that is the end of the gun-shy saga. The incident that started five years later is related, but to me it is not a part of my 15-year quest. That ended in 1991. It will have to wait for another day. I am dog-tired.

MFJ: That was some saga. Thanks for trusting me enough to share it. My stories will not be so dramatic; I think I have had a vanilla life. I notice that you are already starting to think about animals again. See how perceptive I am? Rest well.

❖ 8

MFJ: Good morning. Did you sleep well?

MW: I slept most of the night, which is good for me these days. I am ready to push on, if you do not have any prior needs.

MFJ: The only need I have is hot coffee, and my mug is full. Before you make some bad pun about "my mug", I mean my cup.

MW: My God. I cannot get away from it. You suggested as a light remark that I might, as a bad joke, respond to mug as though you had used it meaning "face" and the association that jumped into my mind was "mug shot." I will just get on with it.

I retired at the end of 1993 and moved to Florida, where we stayed for three years. Before moving, I threw away 25 years of journals I had kept.

MFJ: Why in the world did you do that?

MW: I preferred that my journals not be read by other people. There were no great secrets in the journals, but I wrote them for myself. I did not correct my spelling or hesitate to write down wild hunches or speculations.

In trying to defend my impulsive action, I remember an argument that Anais Ninn had with Henry Miller when they were both living in Paris and discussed their dreams, later fulfilled, of becoming writers. Ninn kept journals and after repeatedly revising them published books in diary form. Miller thought that would be a bad approach for him. He thought it better to just remember his experiences and let the unconscious mind shape them into stories. If I had in the back of my mind someday writing of my experiences, I must have opted to go with Henry Miller rather than with Anais Ninn.

MFJ: It seems I remember you telling me that there was some old-timer on Monday Night Football who liked to say, "I admire his courage and question his judgment."

MW: Meaning you would have chosen differently. Different strokes for different folks. Don Meredith. He had fun on the field and in the announcer's booth. He reminded me of Dizzy Dean. When late in a game, a team scored a touchdown that put it out of reach for the other team, Meredith would start singing, "Turn out the lights; the party's over."

MFJ: Was Meredith your favorite announcer?

MW: Close to it. Even better, though, was radio when you could make up your own pictures of the game. I guess in my mind's eye, nothing will ever come up to the high drama of hearing Bill Stern on radio, calling the Sugar Bowl, in which Georgia had Charlie Trippi and North Carolina had Charlie "Choo Choo" Justice.

Getting back to retirement, I had been very tired and was happy to just walk the beach and learn the game of four-wall racquetball. We had had one-wall racquetball courts at Georgia State in the early days. That is an entirely different game. After a year or so, I was ready to work a little bit and complete the database of animal eye colors that I had worked on, with Polly's help, for many years. We closed it out at 5,620 species. Most were birds. Then, I struggled with choosing the appropriate taxonomies for organization. Next, I devised a five-point scale, which required some subjective judgment. The two-point scale had more face validity, but it lacked a lot of power to discriminate smaller differences in eye color and smaller differences in behavior. [A new, very important, large study of birds (Craig and Hulley 2004), done in South Africa, relates iris pigmentation to a host of other variables. In that study, a three-point scale of eye-darkness was used.] Then, I did the various jobs of scoring each species and calculating average eye darkness for each family, order and class of land vertebrate. The database was now in the final form we wanted to make available to any future researcher who might find it of use.

I then wrote a book-length manuscript of my theoretical notions with descriptive examples in which I tried to integrate information in the database with written reports of animal behavior. I wrote a fairly long introduction of my personal experiences trying to study eye color and behavior. I did the introduction in my best self-effacing style. All in all, I thought I had a pretty good manuscript to take to a publisher. However, as you just pointed out with the Meredith quote, I can have poor judgment about these things.

MFJ: Wait a minute. Don't start getting me in your cross hairs about something that transpired long before we started this process. That is spin up with which I will not put.

MW: You got that line from Winston Churchill. My self-diagnosed personality profile of quirks, knacks and some ability for creative insights also includes the potential to over-estimate or sometimes under-estimate the value of my own work. Before trying to publish the manuscript, I sent it for criticism to a friend. I will call him Bill. He has good editorial skills and has been supportive of me in the past. I knew he was on my side, would pull no punches, nor take cheap shots at me. I had been able to deal with and eventually laugh about strong criticism before without it affecting my motivation. One of my first submissions ever to a journal, a theoretical observation, came back rejected with this brief comment that, as a former writer of military reports, I had to admire, "It is a small point, poorly made."

What I did not know was that my friend who lived in another state was under a lot of stress when I sent him my manuscript. Perhaps for that reason, when he sent it back, his comments seemed to have more of an edge to them than I think he intended. I was not really surprised that he did not like the introduction I had written. What jumped from the page, though, was his comment about it—"The person who wrote the introduction should be shot." It would have been different if he had said, "The person who wrote that introduction should take a long walk on a short pier." I would have said, "That sounds like Bill," laughed it off, taken some of his suggestions, given the manuscript a final revision and moved on to possible publication. All I can think now is that his quip (and it was only that) tapped into one of my old internal programs from childhood about myself. That occurred to me after we started doing these sessions.

What happened back then was that I tore up the introduction and started over. After several tries, I found that I was absolutely blocked. I could not go forward and I could not go back to where I had been before. I knew even then that I had over-reacted to a harmless quip, but knowing that did not help. I gave it up as a lost cause; that was around 1997. In 1999, I had a chance to re-issue in paperback form the 1974 book on eye color, which had been out-of-print for 25 years. I did that and it was re-issued with the title changed to *Eye Color: A Key to Human and Animal Behavior*. After that, I felt a new urge to get the database into print. Rather than go back to the old manuscripts I had written in 1996, I wrote a short introduction that gave a very brief comment on every order of land vertebrate for which we had eye colors for at least 15 species. By including all orders, I demonstrated that the phenomenon was general across the many orders of amphibians, reptiles, mammals, and birds. The bulk of the content is a listing of eye colors for 5620 species. That first unveiling of the database was designed to briefly show the whole picture. It was published as a soft cover book titled *Animal Eye Colors: Yellow-eyed Stalkers, Red-eyed Skulkers, and Black-eyed Speedsters.* I liked the sub-title because it gave a brief summary of the underlying theory.

MFJ: And that I assume brings us to the point several weeks ago when you looked again at the old 1996-97 manuscripts, went to bed feeling blocked and then, woke up with the idea of recording our conversations as a way to comment some more about the database. Do I have it right?

MW: Yes, and something happened in the sessions that modified my plans somewhat. The first was that the process itself, the dialogue technique, seemed to bring to the surface a story I needed to tell and we decided to just go with that for a while. Maybe I don't need to say much more about eye color and my theory. The way I figure it, John, is if you can understand it so easily without any additional explanation, anyone can. Now why do I say things like that when I don't really mean them?

MFJ: Because you are a mean old man.

MW: Don't you know a rhetorical question when you hear one? I would call you a young whippersnapper, if I knew what that meant. I get the impression that there is something on your mind. Is there anything you want to ask or say?

MFJ: You have included in your account of the gun-shy saga synchronicity experiences; that will lose you credibility with some scientists. The other thing is, you have said that as a possible reason or excuse for your writer's block was the quip you received from your friend, Bill, that you "should be shot." Is it logical at all that you would still react that strongly to something that you knew was not intended to be taken literally?

MW: Not at all logical. Remember, though, we are talking about the unconscious mind, which is more concrete and operates with a sort of logic outside the conscious mind. I could give you example after example from my work as a psychotherapist that would still only be examples and not proof. Another off-the-wall example, which is only an example, was my reaction to the Friday night tennis round- robins. You remember I said I had a possible insight the next morning. Friday you and I talked a lot about the gun accident in Pelzer. Then Friday evening I went to play tennis and relax. Riding home I remarked to Polly that unlike my usual experience, I seemed tenser after the tennis than before. What came to me in the twilight zone as I was waking up Saturday morning gave me a clue as to one possible off-the-wall explanation. Before we started, we were told that from now on we would play under "no add" rules. If a game reached deuce, we were to play just one more *sudden death* point to decide the game. Before the evening was over, I had been in many *sudden death* situations, deciding how aggressively I wanted to play the point, and of course I kept hearing that term all night. Consciously I made no special note of it or tied it to what we had been talking about all day. Another thing I remembered while I was waking up was a quip that a woman made following a line call that she had insisted on and then heard that the other three of us on the court had seen it differently. She quipped, with a straight face, "All I can say is that none of you better show up at the hospital where I work as a trauma nurse; I will just let you die." Could all the talk of sudden death and the woman's quip have the effect of making me tense? Could Bill's quip really have tapped into an old unconscious program? It sounds silly in a way, but think about my sleeping bag/watermelon story? How weird is that? And be sure before you put too much stock in the primacy of conscious logic to read that book I suggested, *The User Illusion.* As for synchronicity, who knows? Maybe Jung was on to something and maybe not. If radio waves can be all around us in the air, without us seeing them, why not emotionally-charged brain waves that lead to synchronicity experiences? Have a good evening.

MFJ: You, too. See you tomorrow.

MW: No, John. I think I need a much longer break to just sit in my Brumby rocker and mull over whether or not to publish these things. If I decide to continue, I will give you a call.

MFJ: That is probably wise. I have gotten more than I expected from this and the decision should be entirely up to you. See you at the tennis drills.

MW: I am looking forward to it and more time at the coffee shop. Before we go, though, I need to read something aloud so it will be on the tape:

I WANT TO EMPHASIZE TO ANY READER WHO SUFFERS FROM POST-TRAUMATIC STRESS, THAT IT IS WISE TO ALWAYS BE AWARE OF THE MENTAL HEALTH RESOURCES IN YOUR COMMUNITY.

BREAK

MFJ: I am glad you called and said you wanted to record one more session, but, man, when you take a break, you really take a break. Do you realize that it was seven years ago that we were recording our conversations?

MW: Yes, I know. We were taping in April 2005 and now it is July 2012. The problem is I could never decide how or whether to proceed and, in time, I, more or less, forgot about it. Then, about a week ago, for whatever reason, I picked up the transcripts, started reading and decided that I would publish them if both you and Polly agreed with that decision.

MFJ: You said it was about a week ago. Was that, maybe, the day the temperature got up to 106 degrees? Perhaps that decision was a result of a heat stroke?

MW: Same old John. You make me laugh and that is a good thing.

MFJ: Are we going to talk more about eye color or just stop where we are?

MW: Well, that is one of the things I have been thinking about for the last week and decided that this is a reasonably good place to stop. The eye color pattern is there in nature for anyone who cares to look. I will add an introduction and a bibliography to go with our *conversations* and that will provide some more information. All three of the books I have written about eye color will now be in print. Each provides plenty of unique information not in the other two. With this third one I have brought the terminology up to date. With three books and the bibliography, I think I have provided plenty of leads for future scholars looking for a dissertation topic.

John, I am aware now that there may be readers who have stayed with us to the end. In order to talk to them directly, I am going to read here into the transcript of our *conversations* some suggestions for those who want to follow up on one or the other of our main themes. Then you and I will have to take our parting shots at each other, so give it some thought while I am reading.

Suggestions for the Reader

If you are interested in reading about how changing your ways of thinking can help you to avoid mild depression, read the work of Martin Seligman.

If you are interested in knowing more about how the unconscious mind works and are prepared to be surprised, read first Norretranders's book, *The User Illusion*. After that, if you want to read about a master using his understanding of the unconscious mind to work wonders as a clinical hypnotist, read all you can find about Milton Erickson. Knowledge in this area is increasing so fast that you may, also, want to read something that has been published in the last year or two.

If you are interested in eye color and bird behavior and want to read something written by ornithologists and published in a peer-reviewed journal, start with the Craig and Hulley 2004 article, "Iris colour in passerine birds: Why be bright-eyed?" It was published in the *South African Journal of Science.* Their many findings were very interesting to me and I look forward to seeing a similar large study done with non-passerine birds.

For readers who want some more evidence that eye darkness may serve as a better predictor than race, even when the focus of the study is limited to humans, please read an article about a large study of hypertension, "Race, iris pigmentation, and intraocular pressure," published in 1982 by Hiller in the *American Journal of Epidemiology.*

For readers who want to try one of my other two books on the topic of eye color and behavior, I can guide you. If you are a general reader and want to sample the wide range of behavior, especially human behavior, try my book, *Eye Color: A Key to Human and Animal Behavior.* This book was first published under a different title in 1974 and re-issued in 1999. It reflects the fact that at that time I had just discovered the potential importance of eye color and had gone looking for anything and everything that might be related to eye darkness. Some things may surprise you. I have already mentioned that a passing speculation about eye color and alcoholism was later supported in a large systematic study. The early pilot studies of eye color and athletic performance reported there led to the reactivity hypothesis later confirmed in the laboratory studies of reaction time done at Penn State. The major part of the study of animals reports species differences (ignoring family) within the order of birds that includes eagles, hawks, falcons and vultures.

In that first book, I, perhaps, included too much--everything from marching insects to perception, to social behavior, to biological clocks, to creatures of all ages. By the time the second book, *Animal Eye Colors: Yellow-eyed Stalkers, Red-eyed Skulkers and Black-eyed Speedsters,* was published in 2000, I had narrowed my focus to adult land vertebrates. I want to make it clear again that my hypotheses are not intended to predict behavior of children, fish, insects or individuals of any species including our own.

The main purpose of this second book on eye color and behavior was to publish the database of eye colors (scored for eye darkness) for 5,620 species of land vertebrates. Most were species of birds, but included are eye colors for hundreds of species of mammals, reptiles and amphibians. One study is presented to show that families of birds that feed on the wing are grouped on average eye darkness near the black-eyed end of the scale. By contrast, families that routinely feed on fruit are shown to have average eye darkness scores near the midpoint between yellow and black. The same pattern is shown for bats.

The introduction to the database also includes descriptive statistics and a short summary for every order of land vertebrates for which there is sufficient eye color information in the database. Analyses are included for 2 orders of amphibians, 3 orders of reptiles, 5 orders of mammals and 15 orders of birds. The consistency of the pattern that emerges is a strong indication that the phenomenon generalizes across all classes of land vertebrates.

If there are any readers that I have not sent off somewhere else, you are the ones who just read for fun. I direct you to my book of puzzles, *Aha: A Puzzle Approach to Creative Thinking*. It was first published in 1975. Before reading it, you are encouraged to read the *Wikipedia* entry for "Ditloid." You will read how my book played a role in sparking a fad in Britain for a new form of puzzle.

MW: Back to you, John. Wake up.

MFJ: I just want to say thanks for letting me be a part of the process. I will look forward to more conversations at the coffee shop, and later, if you want to record more of our conversations, I will be more than ready.

MW: Many thanks. As I said, I owe you. I, too, look forward to more informal conversations over coffee.

MFJ: We can continue to spend the mornings or afternoons measuring out our lives with coffee spoons. We may even talk more about animals and how the hunting technique fits the physique. Then in the evenings, we can play tennis.

Is there anything else we should say about animal eye color?

MW: No. Just remember that eye color is related to *habitat, hunting* and *hiding,* as well as *physique* and *technique.* It is time to talk tennis. I am ready to give this tape to Polly. John, after this I am ready for you to just fade away back into your dull life at the coffee shop and tennis court. Sometimes though, I can still hear you talking to me when I am just waking up. To show you my heart is in the right place, you get the last word. I will not respond, but I will take whatever you say into my heart and ponder it.

MFJ: I have already said thanks. I will leave you these words to ponder. Morgan, always remember, you are nothing without me. I see that you are nodding your agreement and laughing. I will hold on to that image and turn off the recorder.

BIBLIOGRAPHY

Asdell, S. A. 1966. *Dog Breeding*. Boston: Little Brown.

Audubon Society. 1977. *Book of wild animals*. New York: Harry N. Abrams.

Austin, L. L. 1961. *Birds of the World*. New York: Golden Press.

Barber, T. X. 1993. *The human nature of birds*. New York: Dorling Kindersley.

Bart, R. S. & Schnall, S. 1973. Eye color in darkly pigmented basal-cell carcinomas and malignant melanomas: an aid in their clinical differentiation. *Archives of Dermatology* 107:206-207.

Bassett, J. F. & Dabbs, J. M., Jr. 2001. Eye color predicts alcohol use in two archival samples. *Personality & Individual Differences* 31:63- 67.

Bauer, E. 1988. *Predators of North America*. Harrisburg, Pennsylvania: Stockpole.

Bellairs, A. 1970. *The life of reptiles*. New York: Universe Books.

Brown, L., and Amadon, D. 1968. *Eagles, Hawks and Falcons of the World*. New York: McGraw-Hill.

Bruce, M. D. 1985. Tyrant Flycatchers and Pittas, in Perrins, C. M., and Middleton, A. L. A. *The Encyclopedia of Birds*. New York: Facts on File 318-323.

Burton, M., and Burton, R. 1984. *Encyclopedia of reptiles, amphibians and other cold-blooded animals*. San Sebastian, Spain: BPC Publishers.

Carney, R.G. 1962. Eye color in atopic dermatitis. *Archives of Dermatology* 85:57-61.

Chapman, J. A., and Feldhamer, G. A. 1982. *Wild Mammals of North America: Biology, Management and Economics*. Baltimore: The Johns Hopkins University Press.

Cogger, H. G., and Zweifel, R. G. 1992. *Reptiles and Amphibians*. New York: Smithmark.

Copeland, R.J., Coleman B., and Rubin, K.H. 1998. Shyness and little boy blue: iris pigmentation, gender, and social wariness in preschoolers. *Developmental Psychobiology* 32:37-44.

Corbet, G. B., and Hill, J. E. 1980. *World List of Mammalian Species*. London: Comstock.

Craig, A.J.F.K. and Hulley, P.E. 2004. Iris colour in passerine birds: Why be bright- eyed? *South African Journal of Science* 100(11&12):584-588.

Cresswell, W. 1996. Surprise as a winter hunting strategy in Sparrowhawks, *Accipiter nisus*, Peregrines, *Falco peregrinus* and Merlins, *F. columbarius*. *Ibis* 138:684-692.

Cronbach, L. J. 1960. *Essentials of Psychological Testing*. New York: Harper and Row.

Dabbs, J. M., and Dabbs, M. G. 2000. *Heroes, Rogues, and Lovers: Testosterone and Behavior*. New York: McGraw-Hill.

Da Costa, E. A., Castro, J.C. and Macedo, M. E. 2008. Iris pigmentation and susceptibility to noise-induced hearing loss. *International Journal of Audiology*, 47(3): 115-118.

Drimmer, F. (Ed.) 1954. *Animal Kingdom,* vol 1. Garden City, New York: Garden City Books.

Eliot, T. S. 1930. *Selected Poems.* San Diego: Harcourt Brace.

Erickson, M. 1964/1980. The surprise and My-Friend-John techniques of hypnosis: Minimal cues and natural field experimentation. In Rossi, E. (Ed) *The Collected Papers of Milton H. Erickson on Hypnosis. 1. The Nature of Hypnosis and Suggestion* (83-107). New York: Irvington.

Ewer, R. F. 1973. *The Carnivores.* Ithaca, New York: Cornel University Press.

Fitzgerald, F. S. 1945. Edited by Edmund Wilson. *The Crack Up.* New York: J. Laughlin.

Forshaw, J. 1991. *Encyclopedia of Birds.* New York: Smithmark.

Friedman, G.D., Selby, J.V., Quesenberry, C.P., Newman, B. and King, M.C. (1990) Eye color and hypertension. *Medical Hypotheses.* 33(3), 201-206.

Frudakis, T., Thomas, M., Gaskin, Z., Venkateswarlu, K., Chandra, K.S., Ginjupalli, S., Gunturi, S., Natrajan, S., Ponnuswamy, V.K. and Ponnuswamy, K.S. 2003. Sequences associated with human iris pigmentation. *Genetics,* 165, 2071-2083.

Gallwey, W.T. 1976 *Inner Tennis: Playing the Game.* New York: Random House.

Goel, N., Terman, M. and Terman, J.S. 2002. Depressive symptomatology differentiates subgroups of patients with seasonal affective disorders. *Depression and Anxiety,* 15, 34-41.

Grandin, T. and Deesing, M.J. 1998. *Behavioral genetics and animal science.* San Diego: Academic Press.

Grzimek, B. (Ed.) 1972-75. *Animal Life Encyclopedia.* New York: Van Nostrand Reinhold.

Hale, B. D., Landers, D. M., Snyder-Bauer, R., and Goggin, N. 1980. Iris Pigmentation and fractionated reaction and reflex time. *Biological Psychology* 10:57-67.

Halliday, R. R., and Adler, K. (Eds.) 1986. *The encyclopedia of reptiles and amphibians.* New York: Facts on File, 78.

Hancock, J., and Kushlan, J. 1984. *The Herons Handbook.* New York: Harper and Row.

Happy, R., and Collins, J. K. 1972. Melanin in the ascending reticular activating system and its possible relationships to autism. *Medical Journal of Australia* 2: 1484-1486.

Harrison, C., and Greensmith, A. 1993. *Birds of the World.* New York: Dorling Kindersley.

Hartman, C. G. 1952. *Possums.* Austin: University of Texas Press.

Higuchi, S., Motohashi, Y., Ishibashi, K., and Maeda, T. 2007. Influence of eye colors of Caucasians and Asians on suppression of melatonin secretion by light. *American Journal of Physiology-Regulatory, Integrative and Comparative Physiology,* 292, R2253-R2256.

Hill, J. E., and Smith, J. D. 1984. *Bats: A Natural History.* Austin: University of Texas Press.

Hiller, R. 1982. Race, iris pigmentation, and intraocular pressure. *American Journal of Epidemiology* 115(5):674-683.

Hood, J. D., Poole, J. P., and Freedman, L. 1976. The influence of eye colour upon temporary threshold shift. *Audiology* 15:449-464.

Jahoda, G. 1971. Retinal pigmentation, illusion susceptibility and space perception. *International Journal of Psychology* 6:199-208.

James, W. 1925. *The Philosophy of William James.* New York: The Modern Library.

James, W.T. 1951. Social organization among dogs of different temperaments, terriers and beagles, reared together. *Journal of Comparative and Physiological Psychology* 44:71-77.

Johnsgard, P. A. 1988. *North American Owls.* Washington: Smithsonian Institution Press.

Kinnear, N. B. 1949. *Whistler's Popular Handbook of Indian Birds* (4th Ed.). Edinburg: Oliver and Boyd.

Kleinstein, R. N., Seitz, M. R., Barton, T. E., and Smith, C. R. 1984. Iris color and hearing loss. *American Journal of Optometry and Physiological Optics* 61(3):145-149.

Krizek, V. 1968. Iris color and composition of urinary stones. *Lancet* 1(7557):1432.

Landers, D. M., Obermeir, G. E., and Patterson, A. H. 1976. Iris pigmentation and reactive motor performance. *Journal of Motor Behavior* 8:171-179.

Landers, D. M., Obermeir, G. E., and Wolf, M. D. 1977. The influence of external stimuli and eye color on reactive motor behavior. In R.W. Christina and D.M. Landers (Eds),

Psychology of Motor Behavior and Sport, 1976, Champaign, ILL: Human Kinetics Publications, 94-112.

Mahut, H. 1958. Breed differences in the dog's emotional behavior. *Canadian Journal of Psychology* 12:35-44.

Marcon, E., and Mongini, M. 1987. *World Encyclopedia of Animals.* New York: Greenwich House.

Mech, L. D. 1970. *The Wolf: the ecology and behavior of an endangered species.* Garden City, New York: The Natural History Press.

Morris, D. *Animalwatching.* New York: Crown. 101-102.

McGinness, J., Corry, P., and Procter, P. 1974. Amorphous semiconductor switching in melanins. *Science* 183:853-855.

Norretranders, T. 1998. *The User Illusion.* New York: Penguin.

Obianwu, H. O., and Rand, M. J. 1965. The relationship between the mydriatic action of ephedrine and the color of the iris. *British Journal of Ophthalmology* 49:264-270.

Owen, D. B. 1962. *Handbook of Statistical Tables.* Massachusetts: Addison-Wesley.

Pearson, T. G. (Ed) 1936. *Birds of America.* Garden City, New York: Doubleday.

Percy, Walker. 1991. Questions they never asked me. In *Signposts in a Strange Land.* New York: Noonday Press. 417a

Perrins, C. (Ed) 2003. *Firefly Encyclopedia of Birds.* Buffalo, New York: Firefly.

Perrins, C. M. 1990. *The Illustrated Encyclopedia of Birds.* New York: Prentice Hall.

Perrins, C. M. and Middleton, A. L. A. 1985. *The Encyclopedia of Birds.* New York: Facts on File, 78.

Petterson, R. 1964. *Silently By Night.* New York: McGraw-Hill.

Pfeffer, P. 1989. *Predators and predation.* New York: Facts on File.

Pirsig, R. 1974. *Zen and the Art of Motorcycle Maintenance.* New York: HarperCollins.

Porter, K. R. 1972. *Herpetology.* Philadelphia: W. B. Sanders.

Prieto, J. G. 1977. Eye color in skin cancer. *International Journal of Dermatology* 16(5):406-407.

Pugnetti, G. 1980. *Simon & Schuster's Guide to Dogs.* New York: Simon and Schuster.

Pugnetti, G. 1983. *Simon & Schuster's Guide to Cats.* New York: Simon and Schuster.

Ridley, M. 2003. *The Agile Gene: How Nature Turns on Nurture.* New York: HarperCollins.

Ripley, S. D. 1988. Revealing the secret lives of the little phantoms of the marshes. *Smithsonian* 19(5):39-45.

Rosenberg, A., and Kagan, J. 1987. Iris pigmentation and behavioral inhibition. *Developmental Psychobiology* 20(4):377-392.

Rubin, K.H. and Both, L. 2004. Iris pigmentation and sociability in childhood: A Re-examination. *Developmental Psychobiology,* 22(7), 717-725.

Rule, L. L. 1967. *Pictorial guide to the mammals of North America.* New York: Thomas Y. Crowell.

Savage, C. 1995. *Bird Brains.* San Francisco: Sierra Club.

Seligman, M. E. P. 1991. *Helplessness: On Depression, Development and Death*. New York: W. H. Freeman.

Shine, R. 1993. *Australian Snakes. A Natural History*. Sydney: Reed Books.

Short, G. B. 1974. Relationship of eye pigmentation to visual acuity. *American Journal of Physical Anthropology* 41(3):503.

Siegel, S. 1956. *Nonparametric Statistics for the Behavioral Sciences*. New York: McGraw-Hill.

Sinclair, S. 1985. *How Animals See: Other visions of our world*. New York: Facts on File.

Smith, J. M., and Misiak, H. 1973. The effect of iris color on critical flicker frequency (CFF). *Journal of General Psychology* 89:91-95.

Sutton, P. R. N. 1959. Association between colour of the iris of the eye and reaction to dental pain. *Nature* 184:122.

Telford, W. H., Hill, W. R., and Hensley, L. 1978. Human eye color and reaction time. *Perceptual and Motor Skills* 47:503-506.

Thomas, G. B., Williams, C. E., and Hoger, N. G. 1981. Some non-auditory correlates of the hearing threshold levels of an aviation noise-exposed population. *Aviation, Space, & Environmental Medicine* 52(9):531-536.

Van Tyne, J., and Berger, A. J. 1965. *Fundamentals of Ornithology*. New York: Wiley.

Verhoef-Verhallen, E.J.J. 2005 *The complete encyclopedia of cats*. Edison, NJ: Chartwell Books.

Walker, H. M. and Lev, J. 1953. *Statistical Inference*. New York: Holt, Rinehart and Winston.

Walls, G. L. 1963. *The vertebrate eye and its adaptive radiation*. New York: Hafner Publishing Company.

Webb, E. J. Campbell, D. T., Schwartz, R. D., and Sechrest, L. 1966. *Unobtrusive Measures: Nonreactive Research in the Social Sciences*. Chicago: Rand McNally.

Weiner, J. 1999. *Time, Love, Memory: A Great Biologist and His Quest for the Origins of Behavior*. New York: Random House.

Whitaker, J. L. 1980. *The Audubon Society field guide to North American mammals*. New York: Alfred A. Knopf.

White, T.M. and Terman, M. 2003, Effect of iris pigmentation and latitude on chronotype and sleep timing. *Chronobiology International* 20, 1193-1195.

Whitfield, P. 1984. *Macmillan illustrated animal encyclopedia*. New York: Macmillan.

Wielgus, A.R. and Sama, T. 2005 Melanin in human irides of different color and age of donor. *Pigment Cell Research* 18(6), 454-464.

Winer, B. J. 1962 *Statistical Principles in Experimental Design*. New York: McGraw-Hill.

Winston, H. D., Lindzey, G., and Conner, J. 1967. Albinism and avoidance learning in mice. *Contemporary Research in Behavioral Genetics* 63:77-81.

Wolf, M. D., and Landers, D. M. 1978 .Eye color and reactivity in motor behavior. In: Landers, D. M., and Cristina, R. W. (Eds.)

Psychology of Motor Behavior and Sport. Champaign, IL: Human Kinetics Publishers, 255-263.

World Atlas of Birds. 1974. New York: Crescent.

Worthy, L. H. (Ed) 1983. *All That Remains: The Traditional Architecture and Historic Engineering Structures, Richard B. Russell Multiple Resource Area, Georgia and South Carolina.* Atlanta: Archeological Services National Park Service, United States Department of the Interior.

Worthy, M. 1975. *Aha: A Puzzle Approach to Creative Thinking.* Chicago: Nelson-Hall.

Worthy, M. 1978. Eye color, size and quick-versus-deliberate behavior of birds. *Perceptual and Motor Skills* 47:60-62.

Worthy, M. 1987. Citation Classic: Self disclosure as an exchange process. *Current Contents* 19:16.

Worthy, M. 1999. *Eye Color: A Key to Human and Animal Behavior.* Lincoln, Nebraska: iUniverse.com [This is a slightly revised version of a book that was published in 1974 and had been out-of-print for 25 years.]

Worthy, M. 2000. *Animal Eye Colors: Yellow-eyed Stalkers, Red-eyed Skulkers & Black-eyed Speedsters.* Lincoln, Nebraska: iUniverse.com.

Worthy, M., Gary, A. L., and Kahn, G. 1969. Self disclosure as an exchange process *Journal of Personality and Social Psychology* 13:59-63.

Worthy, M., and Markle, A. 1970. Racial differences in self-paced versus reactive sports activities. *Journal of Personality and Social Psychology* 16:439-443.

Zhu, G., Evans, D. M., Duffy, D. L., Montgomery, G. W., Medland, S. E., Ewen, K. R., Jewell, M., Liew, Y. W., Hayward, N. K., Strum, R. A., and Martin, N. G. 2004. A Genome Scan for Eye Color in 502 Twin Families: Most variation is due to a QTL on Chromosome 15q. *Twin Research* 7 (2): 197-210.